NASA: A HISTORY OF
THE U.S. CIVIL SPACE PROGRAM

Roger D. Launius

AN ANVIL ORIGINAL

Under the general editorship of
Louis L. Snyder

KRIEGER PUBLISHING COMPANY
Malabar, Florida

1994

Original Edition 1994

Printed and Published by
KRIEGER PUBLISHING COMPANY
KRIEGER DRIVE
MALABAR, FLORIDA 32950

FROM A DECLARATION OF PRINCIPLES JOINTLY ADOPTED BY
A COMMITTEE OF THE AMERICAN BAR ASSOCIATION AND A
COMMITTEE OF PUBLISHERS:

This publication is designed to provide accurate and authoritative information in regard to the subject matter covered. It is sold with the
understanding that the publisher is not engaged in rendering legal,
accounting, or other professional service. If legal advice or other expert
assistance is required, the services of a competent professional person
should be sought.

Printed in the United States of America.

Library of Congress Cataloging-in-Publication Data

Launius, Roger D.
    NASA : a history of the U.S. civil space program / Roger D.
Launius.—Original ed.
    p.   cm.—(The Anvil series)
    "An Anvil original."
    Includes index.
    ISBN 0-89464-727-X
    ISBN 0-89464-878-0 (TR)
    1. United States. National Aeronautics and Space Administration—
History.   2. Space sciences—United States—History.
3. Astronautics—United States—History.   I. Title.
TL521.312.L28   1994
387.8'0973—dc20   93-35977
                                     CIP

10   9   8   7   6   5   4   3

To the memory of Eugene M. Emme (1919–1985),
the First NASA Historian

CONTENTS

PART I

PART II Readings

PREFACE AND ACKNOWLEDGMENTS

The later twentieth century has been characterized in many ways but humanity's movement beyond Earth, with both machines and people, may be its most singular development. Even at this time, only a little more than a generation after the first orbital flights, the compelling nature of the space age and the activity that it has engendered on the part of many people and government organizations makes the U.S. civil space program a significant area of investigation. At the same time, questions about why it took the shape it did; which individuals and organizations were involved; what factors influenced particular choices of scientific objectives and technologies to be used; and the political, economic, managerial, international, and cultural contexts in which the events of the space age have unfolded are especially pertinent as the U.S. government seeks to come to grips with future policies and programs.

This volume is designed as a short historical synthesis of the U.S. civil space program since the first experimentation with rocket technology in the early part of the twentieth century through NASA's attempts to build a permanently-occupied space station in the 1990s. As an entirely twentieth century phenomenon, space exploration's history is relatively short but it is nonetheless important in the modern development of the United States. This account is only an introduction a rich subject. It treats both the significant and the more mundane aspects of the space program with the intention of creating a portrait of an endeavor that has repercussions across a broad spectrum. The space program has evolved since the 1950s within and from a diverse combination of interests within a Cold War context of competition with the Soviet Union. These interests were politically, foreign policy, technologically, economically, and scientifically oriented. While the Cold War has subsided and many of the individual interests have shifted over time the basic combination of elements making up the base of the space program have remained. This change over time is a key element in this study.

As in all of the Anvil series of historical books, this one

contains two parts. The first part is a brief narrative history. It can do no more than highlight major themes and deal with a few specifics as examples. It begins with a chapter discussing in broad terms the development of rocket technology in the early twentieth century, focusing on the pioneering work of Robert H. Goddard in the United States and the efforts of other engineers to build the first practical rockets during World War II. Several chapters on the Cold War competition with the Soviet Union and the response of the U.S. in building a civil space program follow, as well as a discussion of the environment and activities of the space program in the 1980s and early 1990s. The second part of this book is documentary in nature, providing excerpts from 25 key documents in U.S. space program history. These, too, are only representative of a much richer set of documentary materials available for use by those wishing to understand the development of the civil space program.

Whenever historians take on a project of historical synthesis such as this, they stand squarely on the shoulders of earlier investigators and incur a good many intellectual debts.

I would like to acknowledge the assistance of several individuals who aided in the preparation of this overview of U.S. space program history. Lee D. Saegesser, NASA History Office Archivist, was instrumental in obtaining the documents included in the book and in providing other archival services; J.D. Hunley, assistant historian in the NASA History Office, edited and critiqued the text; Patricia Shephard helped with proofreading and compilation; the staffs of the NASA Headquarters Library and the Scientific and Technical Information Program provided assistance in locating materials; and archivists at various presidential libraries and the National Archives and Records Administration aided with research efforts. Several scholars read portions of the manuscript or talked with me about the project, in the process helping me more than they could ever know. These included Roger L. Bilstein, Tom D. Crouch, Robert Dallek, Virginia P. Dawson, Duane Day, Henry C. Dethloff, Andrew Dunar, Linda Neumann Ezell, Robert H. Ferrell, Adam L. Gruen, R. Cargill Hall, Joan Hoff, Karl Hufbauer, Sylvia K. Kraemer, W. Henry Lambright, John M. Logsdon, Howard E. McCurdy, Pamela E. Mack, John E.

Naugle, Allan E. Needell, Arthur L. Norberg, Lyn Ragsdale, Joseph N. Tatarewicz, Craig B. Waff, Stephen Waring, and Mike Wright. All of these people would disagree with some of the areas I chose to emphasize, with many of my conclusions, and with a few of my document selections, but such is both the boon and the bane of historical inquiry.

PART I

NASA: A HISTORY OF
THE U.S. CIVIL SPACE PROGRAM

CHAPTER 1

PRELUDE TO THE SPACE PROGRAM

UNDERSTANDING THE OBSERVABLE UNIVERSE.
Curiosity about the universe and other worlds has been one of
the few constants in the history of humankind. Prior to the
twentieth century, however, there was little opportunity to explore the universe except in fiction and through astronomical
observations. These early efforts led to the compilation of a
body of knowledge that inspired as well as informed the efforts
of scientists and engineers who began to think about applying
rocket technology to the challenge of spaceflight in the early
part of the twentieth century. These individuals were essentially
the first spaceflight pioneers, translating centuries of dreams
and observations into a reality that matched in some measure the
expectations of the public that watched and the governments that
supported their efforts.

Ancient societies watched the heavens, erecting great observatories to chart the paths of the Sun, Moon, planets, and stars.
Astronomy became wrapped up in their religions, science, and
philosophy. The prehistoric people who built Stonehenge in
England apparently used their observations of celestial bodies to
chart planting seasons and measure other events, assigning this
study a religious significance as well. To the ancient Egyptians,
the Milky Way was a "heavenly Nile" that helped explain for
them some of the seemingly mysterious forces of nature and
build a cosmology of the immortality of the human soul.
Astronomers in Babylon about 700 B.C.E. charted the paths of
several planets and compiled observations of fixed stars. Later,
around 400 B.C.E. the Babylonians devised the zodiac, the first
mechanism to divide the year into lunar periods and to assign
significance to a person's date of birth in foretelling the future.
In what became the Americas the ancient Incan and Aztec
cultures also built astronomical observatories. North American
Indians observed the supernova of 1054 which created the Crab
Nebula—not seen in Europe.

By 150 C.E. Ptolemy of Alexandria—at once a mathematician, geographer, and astronomer—had systematized a large amount of ancient information, and in some cases misinformation, about the universe. Ptolemy based his understanding on classical conceptions of the universe inherited from the Greeks. His great synthesis, *Megiste Syntaxis*, sometimes called the *Almagest*, placed the Earth at the center of the universe, and the Sun, stars, and planets were embedded as jewels in a setting of spheres circling around it. The Roman philosopher and statesman, Cicero, summarized the Ptolemaic belief this way:

The Universe consists of nine circles, or rather of nine moving globes. The outersphere is that of the heavens, which embraces all the others and under which the stars are fixed. Underneath this, seven globes rotate in the opposite direction from that of the heavens. The first circle carries the star known to men as Saturn; the second carries Jupiter, benevolent and propitious to humanity; then comes Mars, gleaming red and hateful; below this, occupying the middle region, shines the sun, the chief, prince and regulator of the other celestial bodies, and the soul of the world which is illuminated and filled by the light of its immense globe. After it, like two companions, come Venus and Mercury. Finally, the lowest orb is occupied by the Moon, which borrows its light from the sun. Below this last celestial circle there is nothing by what is mortal and perishable except for the minds granted by the gods to the human race. Above the Moon, all things are eternal. Our Earth, placed at the center of the world, and remote from the heavens on all sides, stays motionless and all heavy bodies are impelled towards it by their own weight. The motion of the spheres creates a harmony formed out of their unequal but well-proportioned intervals, combining various bass and treble notes into a melodious concert.

His position accepted the geometric and static harmony of the universe and placed man squarely at its center, a view not inconsistent with Christian religious ideals about humanity's special relationship with God.

While this world view was modified to some degree during the following centuries, it did not begin to change appreciably until the sixteenth century. The Polish astronomer-mathematician Nicholas Copernicus (1473–1543) saw that the irregular flight of some planets could not be explained using the Ptolemaic conception of the universe, so he placed the Earth in orbit around the Sun. He was, however, circumspect in his public statements about this finding. Others followed his observation to its logical conclusion, most importantly Galileo Galilei (1564–1642), who used the newly invented telescope to show that the Earth and the planets revolved around the Sun in a dynamic and ever-changing universe. In 1616 Galileo was brought before the Inquisition in Rome and his ideas were declared heretical because they challenged more than 1,000 years of Christian tradition about humans being the center of the universe. He received no punishment by agreeing not to teach these ideas, but in 1632 Galileo was brought to trial again for violating his previous compact and was forced into retirement at his home in Florence. He was also compelled to recant his belief that the Earth revolves around the Sun. His legendary response was reported only later, "*E pur si muove*," ("And it does so move").

Regardless of the narrow position of the church, others took up the cause after Galileo. An Englishman, Isaac Newton (1642–1727), was by far the most important, laying a scientific foundation that remained unshaken for two centuries. He formulated the three laws of motion that have been central to the development of space vehicles and placed both astronomy and physics on a firm scientific foundation. He established physical laws governing all matter that could be both mathematically theorized and scientifically observed. Newton's observations on motion suggested that the universe was in constant flux and that motion in a predictable pattern was the natural condition. Universal gravitation, Newton argued, accounted for the physical actions of celestial bodies. In particular, he demonstrated that the attraction of the Sun on a planet was directly proportional to the product of the two masses and inversely proportional to the square of the distance separating them. Newton's ideas, so critical to the development of spaceflight, became the established method of explaining the universe during his lifetime and remained so until the twentieth century. During the intervening period the history of astronomy and physics was mainly a story of working out a Newtonian "system of the world."

THE DREAM OF SPACEFLIGHT. The more than two millennia of searching for answers about the universe exemplified an overriding interest on the part of humanity to learn about its physical surroundings. When Galileo first broadcast his findings about the solar system, it sparked a flood of speculation about lunar flight. Johann Kepler (1571–1630), himself a pathbreaking astronomer, posthumously published a novel, *Somnium* (*Dream*) (1634), that recounted a supernatural voyage to the Moon in which the visitors encountered serpentine creatures. He also included much scientific information in the book, speculating on the difficulties of overcoming the Earth's gravitational field, the nature of the elliptical paths of planets, the problems of maintaining life in the vacuum of space, and the geographical features of the Moon.

Other writings sparked by the invention of the telescope and the success of *Somnium* also described fictional trips into space. Savinien Cyrano de Bergerac (1610–1655), for example, wrote *Histoire comique des états et empire de la lune* (*Voyage to the Moon*) (1649) describing several attempts to travel to the Moon. In it Cyrano was catapulted to the Moon by explosives. Thus, he became the first flyer in fiction to reach the Moon by means of rocket thrust, a premonition of Newton's third law of gravity about every action having an equal and opposite reaction. Once on the Moon, the character in this novel had several adventures, and later in the book the primary character also journeyed to the Sun.

Other writers picked up the science fiction format and used it to discuss the possibilities of space travel in the years that followed. For example, Edward Everett Hale (1822–1909), a New England writer and social critic, published in 1869 a short story in the *Atlantic Monthly* entitled "The Brick Moon." The first known proposal for an orbital satellite around the Earth, Hale described how a satellite in polar orbit could be used as a navigational aid to ocean-going vessels.

For you see that if, by good luck, there were a ring like Saturn's which stretched around the world, above Greenwich and the meridian of Greenwich, . . . anyone who wanted to measure his longitude or distance from Greenwich would look out of his window and see how high this ring was above his horizon. . . . So if we only had a ring like that . . . vertical to the plane of the equator, as the brass ring of an artificial globe goes, only far higher in proportion . . . we could calculate the longitude.

The heroes of the story substituted a brick moon for this ring—brick because it could withstand fire—and hurled it into orbit 5,000 miles above the Earth. The remainder of the story described the exploits of the inhabitants of the brick moon.

Perhaps the most important development in the literary consideration of space travel came following the publication of work by Italian astronomer Giovanni Schiaparelli (1835–1910) in 1877 concerning the possibility of canals on Mars. He and others concluded that the features that he saw on Mars and called canals were the work of intelligent life. This was a startling observation because it meant that science had now validated the speculations of some fiction writers, lending credibility to their claims. Moreover, other scientists sought to explore these ideas, and in the United States Percival Lowell (1855–1916) built what became the Lowell Observatory near Flagstaff, Arizona, to study the planets. In 1906 he published *Mars and Its Canals*, which argued that Mars had once been a watery planet and that the topographical features known as canals had been built by intelligent beings. Over the course of the next forty years a steady stream of other works took as a departure point these observations about Mars.

While most of these writings were not scientifically valid, that became less true as time passed and more modern science fiction writers appeared such as Jules Verne (1828–1905) and H. G. Wells (1866–1946). Both were well aware of the scientific underpinnings for spaceflight and their speculations reflected reasonably well what was known at the time about its problems and the nature of other worlds. They incorporated into their novels a much more sophisticated understanding of the realities of space than had been available before. Their space vehicles became enclosed capsules powered by electricity, and they possessed some aerodynamic soundness. Most of Wells's and Verne's concepts stood up under some (but not too rigorous) scientific scrutiny. For example, in 1865 Verne published *De la terre à la lune* (*From the Earth to the Moon*), containing scientific principles that were very accurate for the period. It described the problems of building a vehicle and launch mechanism to visit the Moon. Using a 900-foot-long cannon, Verne's characters were shot into space at the end of the book. Verne picked up the story in a second novel, *Autour de la lune* (*Around the Moon*), describing a lunar orbital flight, but he did not allow his characters actually to land. Wells published *War of the Worlds* in 1897 and *The First Men in the Moon* immediately thereafter. Both used sound scientific principles to describe space travel and encounters with aliens.

The War of the Worlds, furthermore, played upon a theme in space exploration that had been present for many centuries and would continue to be throughout the twentieth century, the cyclical fascination and terror of humanity about contact with alien species. An excitement about the prospect that humanity is not alone in the universe, that contact is possible, and that both cultures might be made richer through interaction has been a persistent theme for advocates of the exploration of space. There is so much that could be learned from more advanced civilizations from other worlds, some claimed. Science fiction expressed well this positive image of contact with aliens—for example, in three novels by C. S. Lewis (1898–1963), *Out of the Silent Planet* (1938), *Perelandra* (1943), and *That Hideous Strength* (1945). At the same time, there has long been a fear that an alien civilization might attack Earth and either enslave or destroy humanity. In *War of the Worlds* the Earth was attacked by invaders from Mars, who eventually were only defeated by terrestrial bacteria harmless to humans but deadly to an alien without generations of built-up immunity. These stories provided some of the inspiration for many scientists and engineers who developed modern rocketry.

THE TECHNOLOGY OF ROCKETS. Until the twentieth century, study about the universe and speculation about the nature of space flight were not closely related to the technical developments that led to rocket propulsion. But a merging of these ideas had to take place before the space age could truly begin. The rocket is a reaction device, based on Newton's Third Law of Motion. Without explicitly understanding this law, humanity has recognized the rocket's practicality for centuries. Although it is unclear who first invented rockets, many investigators link the development of the first crude rockets with the discovery of gunpowder. The Chinese, moreover, had been using gunpowder for some 1,800 years. The first firecrackers seemed to have appeared about the first two centuries after the beginning of the common era, and the Chinese were using rockets in warfare at least by the time of Genghis Khan (ca. 1155–1227). Not long thereafter the use of rockets in warfare began to spread to the West, and was in use by at least the time of the German Albertus Magnus (1193–1290), who gave a recipe for making gunpowder and wrapping it "in a skin of paper for flying or for making thunder" in *De Mirabilibus Mundi* (*On the Wonders of the World*). By the time of Konrad Kyser von Eichstadt, who wrote *Bellfortis* in 1405, the use of rockets in military operations was reasonably well known in Europe.

The use of gunpowder rockets was refined through the first part of the nineteenth century. Essentially, the military application for rocketry—and there was little other at the time—was as a type of artillery. Sir William Congreve (1772–1828) carried rocket technology as far as it was to go for another century, developing incendiary barrage missiles for the British military that could be fired from either land or sea. They were used with effect against the United States in the War of 1812; it was probably Congreve's weapons that Francis Scott Key wrote about in the "Star Spangled Banner" while imprisoned on a British warship during the bombardment of Fort McHenry at Baltimore. The military use of the rocket was soon outmoded in the nineteenth century by developments making artillery both more accurate and more destructive, but new uses for rockets were found in other industries such as whaling and for sea-going shipping where rocket-powered harpoons and rescue lines began to be employed.

PROGENITORS OF THE SPACE AGE. While the technology of rocketry was moving forward on other fronts, some individuals began to advocate their use for space travel. There were three great pioneering figures in this category. Collectively, they were the progenitors of the modern space age. The earliest was the Russian theoretician Konstantin Eduardovich Tsiolkovsky (1857–1935), who had been inspired by the science fiction of Verne and Wells. An obscure schoolteacher in a remote part of Tsarist Russia in 1898, he submitted for publication to the Russian journal, *Nauchnoye Obozreniye* (*Science Review*), a work based upon years of calculations that laid out many of the principles of modern spaceflight. His article was not published until 1903, but it opened the door to future writings on the subject. In it Tsiolkovsky described in depth the use of rockets for launching orbital space ships. Tsiolkovsky continued to theorize on the subject of spaceflight until his death. Significantly, he never had the resources—or perhaps the inclination—to experiment with rockets himself. His theoretical work, however, influenced later rocketeers and served as the foundation of the Soviet space program.

A second rocket pioneer was Hermann Oberth (1894–1989), by birth a Transylvanian but by nationality a German. Oberth began studying the nature of spaceflight about the time of World War I and published his classic study, *Die Rakete zu den Planetenraumen* (*Rockets in Planetary Space*) in 1923. It was a thorough discussion of almost every phase of rocket travel. He posited that a rocket could travel in the void of space. He noted that with the proper velocity a rocket could launch a payload into orbit around the Earth, and to accomplish this goal he reviewed several propellant mixtures to increase speed. He also designed a rocket that he believed had the capability to reach the upper atmosphere by using a combination of alcohol and hydrogen as fuel. Among his proteges was Wernher von Braun (1912–1977), the senior member of the rocket team that built NASA's Saturn launch vehicle for the actual trips to the Moon in the 1960s.

Finally, the American Robert H. Goddard (1882–1945) pioneered in the use of rockets for spaceflight. Motivated by reading Verne's science fiction as a boy, Goddard early became excited by the possibility of exploring space. At his high school oration in 1904 he summarized his future life's work: "It is difficult to say what is impossible, for the dream of yesterday is the hope of today and the reality of tomorrow." As a young physics graduate student he worked on rocket propulsion and actually registered two patents in 1914, one for a rocket using solid and liquid fuel and the other for a multistage rocket. After working for the military as a civilian in World War I, Goddard became a professor of physics at Clark University in Worcester,

Massachusetts. There he published his classic study, *A Method of Reaching Extreme Altitudes*, in 1919. His ideas, however, were ridiculed by the popular press.

The negative press Goddard received made him secretive and reclusive—even paranoid—about security for his experiments. It did not, however, stop his work, and he eventually registered 214 patents on various components of rockets. In time he concentrated on the design of a liquid fueled rocket, the first such development, and the attendant fuel pumps, motors, and control components. On 16 March 1926 he launched near Auburn, Massachusetts, his first rocket, a liquid oxygen and gasoline vehicle that rose 184 feet in 2.5 seconds. It heralded the modern age of rocketry and set the stage for all later space travel. He continued to experiment with rockets and fuels for the next several years. A spectacular launch took place on 17 July 1929 when he flew the first instrument payload—an aneroid barometer, a thermometer, and a camera—to record the readings. The launch failed; after rising about 90 feet the rocket turned and struck the ground 171 feet away. It caused such a fire that neighbors complained to the state fire marshal and Goddard was enjoined from making further tests in Massachusetts.

As time passed Goddard became more secretive and paranoid about his work, not wanting to share information about it with others for fear that they would either ridicule or steal it from him. His ability to shroud his research in mystery was greatly enhanced by Charles A. Lindbergh, fresh from his trans-Atlantic solo flight, who helped Goddard obtain a series of grants from the Guggenheim Fund fostering aeronautical activities. This enabled him to acquire a large tract of desolate land near Roswell, New Mexico, and to set up an independent laboratory to conduct rocket experiments far away from anyone else. Between 1930 and 1941 Goddard carried out ever more ambitious testing of rocket components in the relative isolation of New Mexico. The culmination of this effort was a successful launch of a rocket to an altitude of between 8,000 and 9,000 feet in 1937. With World War II starting, in late 1941 Goddard began work as a civilian with the Navy and spent the war developing a Jet-Assisted Takeoff (JATO) rocket to shorten the distance required for heavy aircraft launches. Some of this work helped the development of the throttlable Curtiss-Wright XLR25-CW-1 rocket engine that powered the Bell X-2 in transonic flight research. He did not live to see this, however, for he died in Annapolis, Maryland, on 10 August 1945. Government recognition of Goddard's work came in 1960 when the Department of Defense and the National Aeronautics and Space Administration (NASA) awarded his estate $1 million for the use of his patents.

THE ROCKET AND MODERN WAR. Although the work of Goddard and others was pathbreaking, only World War II truly altered the course of rocket development. Prior to that conflict technological progress had been erratic. The war forced nations to focus attention on applicable technology and to fund research and development. This research was oriented, however, toward the development of rocket-borne weapons rather than rockets for space exploration and other peaceful purposes. This policy would remain the case even after the war, as competing nations perceived and supported advances in space technology because of their military potential and the national prestige associated with them. The role of the Department of Defense and the function of NASA as a civilian space agency have been inextricably related ever since.

During World War II many combatants were involved in developing some type of rocket technology. As an example, the Soviet Union fielded the "Katusha," a solid fueled rocket six feet in length and carrying almost fifty pounds of explosives that could be fired from either a ground- or truck-mounted launcher. The United States began in earnest in 1943 to develop a rocket capability, and several efforts were aimed in that direction. One of the most significant was at the Jet Propulsion Laboratory (JPL) in Pasadena, California, where a team under the brilliant Hungarian scientist, Dr. Theodore von Kármán (1881–1963), began developing the WAC Corporal, which became a significant launch vehicle in postwar rocket research. Others built various types of hand-held antitank and antiaircraft rockets as well as the JATO rockets already mentioned. Germany, however, had the greatest success in developing an operational missile capability. Two weapons in particular were put into use, the V-1 "Buzz Bomb" and the V-2 rocket. The V-1, first used in June 1944, had one substantial weakness; it was relatively slow with a top speed of less than 400 miles per hour. This made it possible for allied pilots and antiaircraft operators to destroy it. Of the more than 8,000 of these weapons launched, over half were destroyed before reaching their targets. But the "Buzz Bombs" that reached London exacted a toll.

While the V-1 was essentially an air-breathing cruise missile, the second German weapon was the first true ballistic missile. The brainchild of Wernher von Braun's rocket team operating at a secret laboratory at Peenemünde on the Baltic coast, this rocket was the immediate antecedent of those used in the U.S. space program. A liquid propellant missile extending some 46 feet in length and weighing 27,000 pounds the V-2 flew at speeds in excess of 3,500 miles per hour and delivered a 2,200 pound warhead. First flown in October 1942, it was employed against targets in Europe beginning in September 1944, and by the end of the war 1,155 had been fired against England and another 1,675 had been launched against Antwerp and other continental targets. The guidance system for these missiles was imperfect and many did not reach their targets, but they struck without warning and there was no defense against them. As a result the V-2s had a terror factor far beyond their capabilities.

That Germany had been so astoundingly successful in developing a ballistic missile while the other combatants had not done so was no accident, and it was in no small measure the result of personalities involved in the research. Although in many ways before 1941 the United States had led the world in rocket technology, chiefly because of the work of Goddard, his reclusive personality had failed to spark the support of either other scientists or the government. On the other hand, the effervescent Oberth had courted his scientific colleagues and those in positions of government responsibility in Germany. For instance, as early as 1929 Oberth had helped kindle the fires of rocketry's promise in Walter Dornberger, later the military commander of the German rocket program. No similar level of salesmanship took place in any other nation. Popular support was therefore lacking, and Germany was able to capitalize on this with the V-2 development during the war.

POSTWAR ROCKETRY. Clearly the technology employed in both of these weapons was worthy of American study and as the war was winding down, U.S. forces brought captured

V-1s and V-2s back for study. Along with them—as part of a secret military operation called Project Paperclip—came many of the scientists and engineers who had developed these weapons, most notably von Braun and several associates from Peenemünde who made a point of surrendering to the Americans so they might continue their work after the war. They were installed at Fort Bliss in El Paso, Texas, and launch facilities for the V-2 test program were set up at the nearby White Sands Proving Ground in New Mexico. Later, in 1950 von Braun's band of over 100 people moved to the Redstone Arsenal near Huntsville, Alabama, to concentrate on the development of a new missile for the Army. Meantime, in Project Hermes, the first successful American test firing of the captured V-2s took place at White Sands on 16 April 1946. Between 1946 and 1951, 67 captured V-2s were test launched.

They were involved in an experimental program to learn more both about the rocket technology and about the upper altitudes, the latter through instruments placed in the warhead section. To coordinate these experiments, in January 1947 the War Department established an Upper Atmosphere Research Panel, and although its name and scope of responsibilities changed periodically during the next several years, it continued to coordinate these activities until the birth of NASA in 1958.

Throughout the late 1940s and early 1950s technicians conducted ever more demanding test flights and explored the idea of orbiting spacecraft. (*See Reading No. 1.*) One of the most important series was Project Bumper, which mated a smaller Army WAC Corporal missile, produced at JPL, as a second stage on a V-2 to obtain data on both high altitudes and the principles of two-stage rockets. The only fully successful launch took place on 24 February 1949, when the V-2/WAC Corporal reached an altitude of 244 miles and a velocity of 5,150 miles per hour. Additionally, the Naval Research Laboratory was involved in sounding rocket research—nonorbital instrument launches using a Viking launch vehicle built by the Glenn L. Martin Co. Viking 1 launched from White Sands on 3 May 1949, while the twelfth and last Viking took off on 4 February 1955. The program uncovered significant scientific information about the upper atmosphere and took impressive high-altitude photographs of earth.

The Army also developed the Redstone rocket during this same period, a missile capable of sending a small warhead a maximum of 500 miles. Built under the direction of von Braun and his German rocket team in the early 1950s, the Redstone took many features from the V-2, added an engine from a Navaho test missile, and incorporated some of the components from other rocket test programs in its electronic components. The first Redstone was launched from Cape Canaveral, Florida, on 20 August 1953. An additional 36 Redstone launches took place through 1958, notably one on 8 August 1957 testing blunt-body shapes and ablative materials use to combat the rigors of superheating during reentry. This rocket led to the development of the Jupiter C, an intermediate-range ballistic missile that could deliver a nuclear warhead to a site after a nonorbital flight into space. Its capability for this mission was tested on 16 May 1958 when combat-ready troops first fired the rocket. The missile was placed on active service with American units in Germany the next month, and served until 1963.

THE BALLISTIC MISSILE PROGRAM. During this same era all the U.S. armed services worked toward the fielding of intercontinental ballistic missiles (ICBM) that could deliver warheads to targets half a world away. Competition was keen for the development of an intercontinental ballistic missile. The first generation ICBM was the Atlas, followed quickly by the Titan missile in the early 1960s. To consolidate military efforts, Secretary of Defense Charles E. Wilson (1886–1972) issued a decision on 26 November 1956 that assigned responsibility for land-based ICBM systems to the Air Force and sea-launched missiles to the Navy. The Navy immediately stepped up work for the development of the submarine-launched Polaris ICBM, which first successfully fired in January 1960.

The U.S. Air Force did the same with land-based ICBMs. The Atlas received high priority from the White House and hard-driving management from Brigadier General Bernard A. Schriever (1900–   ), a flamboyant and intense Air Force leader. The first Atlas rocket test fired on 11 June 1955, and a later version became operational in 1959. These systems were followed in quick succession by the Titan ICBM and the Thor intermediate-range ballistic missile. By the latter 1950s, therefore, rocket technology had developed sufficiently for the creation of a viable ballistic missile capability. It effectively shrank the size of the globe, and the United States—which had always before been protected from outside attack by two massive oceans—could no longer rely on natural defensive boundaries.

The United States's own ICBM capability, additionally, signaled for the rest of the world that the nation could project military might anywhere in the world. More important for the theme of this volume, this military capability could be tapped for the projection of a human and robotic presence into space. The dreams of Verne and Wells were combined with the pioneering rocketry work of Goddard and Oberth and Tsiolkovsky to create the reality of the space age that was on the verge of dawning.

CHAPTER 2

THE SPUTNIK CRISIS

SPACE AND THE COLD WAR. The United States was not the only power in the immediate post-World War II era at work on the development of rockets capable of space travel. The history of space and rocketry during the twenty years after World War II was almost entirely propelled by the rivalry between the United States and the Soviet Union, as the two great superpowers engaged in a "cold war" over the ideologies and allegiances of the nonaligned nations of the world. The vast resources, enormous land areas, and technological capabilities of these two nations catapulted them into the forefront of space exploration, and the intense competition between them ensured that they would dedicate significant resources to the effort. Both saw space as a "new high ground" that must not be abandoned to the other, and its offensive and defensive potential exploited, or at least neutralized so that the other did not exploit it. It was this rivalry that prompted the development of a formal U.S. civil space program. Moreover, the intensity of the rivalry that fueled the U.S. space program was the direct result of a tiny beeping satellite known as Sputnik I, orbited by the Soviet Union on 4 October 1957. This was, literally, a shot heard around the world and nothing has been the same since.

The background of the Sputnik crisis dates from the end of World War II in 1945, when General Henry H. Arnold (1886–1950), commander of the Army Air Forces, confidently assured the Secretary of War, Robert P. Patterson (1891–1952), that the United States would soon be able to field ballistic missiles capable of delivering nuclear weapons half-a-world away and "space ships capable of operating outside the atmosphere." The fear of Soviet aggression sparked research to make Arnold's suggestions a reality during the latter 1940s and early 1950s and led to the development of the offensive intercontinental ballistic missile (ICBM) program and the defensive Distant Early Warning (DEW) Line and other radar systems to detect foreign threats from both conventional bombers and spacecraft.

These predecessors to the formal U.S. space program were both civil and military in scope. Essentially two sides of the same coin, the space exploration efforts of the early years were motivated by the cold war and fostered by the efforts of one key American president, Dwight D. Eisenhower (1890–1969). Elected in November 1952, Eisenhower had been a career Army officer and Supreme Allied Commander of forces in Europe during 1942–1945. Following the war he had remained in Europe as the chief official of the Allied occupation, and he had been instrumental in putting wheels under western Europe's postwar rebuilding effort and the North Atlantic Treaty Organization (NATO) defensive agreement. When elected to the presidency on the Republican ticket in 1952, Eisenhower had many goals but one of those was ensuring the safety and prerogatives of the United States in the face of what he considered an aggressive Soviet Union.

Eisenhower's perspectives on the intentions of the Soviet Union were prompted by several factors, all of which made sense at the time to most Americans.

1. He and most of his advisors were opponents of the communist ideology and economic system not only on philosophical grounds but also because it avowed both by edict and action the conquest of the world's established governments.

2. The empire-building of the Soviet Union at the end of World War II by incorporating Poland, the Baltic republics, eastern Germany, most of the rest of Eastern Europe, some of the Balkans, and parts of Japanese-occupied Asia into its sphere of influence reinforced the sense of immediacy and threat from the Soviet Union that Eisenhower and many of his advisors already philosophically acknowledged.

3. The Soviet attempt to pinch off Berlin during a land blockade in 1948–1949, only relieved by a massive airlift of food and other supplies to a city under siege, further reinforced the perception of an expansionistic Soviet Union.

4. The overthrow of nationalist China under Chiang Kai-shek (1887–1976) by a communist force under the leadership of Mao Zedong (1893–1976) in 1949 greatly enhanced the strength and influence of the communist world.

5. Communist China's support of communist North Korea in an internecine struggle with South Korea, supported by the

United States, between 1950 and 1953 also prompted Eisenhower to take a strong countering position.

Eisenhower believed for these reasons, as well as others of a more subtle nature, that the world was not a safe place for the United States to engage in its traditional foreign policy objectives of isolationism and world trade.

To combat the Soviet Union, Eisenhower supported the development of ICBMs as a deterrent to nuclear attack and of reconnaissance satellites as a means of learning about potentially aggressive actions. The safety from surprise attack promised by reconnaissance satellites was an especially attractive feature for Eisenhower and leaders of his generation because they remembered well the Japanese attack at Pearl Harbor on 7 December 1941 and were committed to never being caught off guard again. At a meeting of key scientific advisors on 27 March 1954 to discuss the use of space for military purposes, Eisenhower warned that "Modern weapons had made it easier for a hostile nation with a closed society to plan an attack in secrecy and thus gain an advantage denied to the nation with an open society." Reconnaissance satellites were a counter to this threat. Issues of national security, therefore, prompted most of the Eisenhower administration's interest in the space program during the 1950s.

RECONNAISSANCE SATELLITES AND FREE ACCESS TO SPACE. While the development of rocket technology was fostered through the military ICBM program, without the pairing of those rockets with orbiting reconnaissance satellites the space program would not have developed as it did. In a critical document, "Meeting the Threat of Surprise Attack," issued on 14 February 1955, defense officials recommended beginning immediately a program to develop a small satellite that would operate at extreme altitudes over foreign airspace. The report also raised the question of the international law of territorial waters and airspace, in which individual nations controlled those territories as if they were their own soil. That international custom—not universally accepted—allowed nations legally to board and confiscate vessels within territorial waters near their coastlines and to force down aircraft flying in their territorial airspace. But space was a territory not defined as yet, and the U.S. position called for it to be recognized as free territory not subject to the normal confines of territorial limits. An opposite position, however, argued for the extension of territorial limits into space above a nation into infinity.

"Freedom of space" was an extremely significant issue for those concerned with orbiting satellites, because the imposition of territorial prerogatives outside the atmosphere could legally restrict any nation from orbiting satellites without the permission of nations that might be overflown. Because the United States, with its superior technological capability, was in a position to capitalize on "freedom of space," it favored an open position. Most other nations, on the other hand, had little capability to launch satellites and therefore had little interest in establishing a free access policy that allowed the United States to orbit reconnaissance or any other types of satellites. The Soviet Union was a wild card in this debate. While it had the potential to launch satellites, it was thought technologically inferior to the United States and might clamor for a closed access position if the United States orbited a satellite first. From the perspective of the Eisenhower administration, which was committed to development of an orbital reconnaissance capability as a national defense initiative, an international agreement to ban satellites from overflying national borders without the individual nation's permission was unacceptable.

Eisenhower tried to obtain a "freedom of space" decision on 21 July 1955 when he attended a U.S./U.S.S.R. summit in Geneva, Switzerland. The absence of trust among states and the presence of "terrible weapons," he argued, provoked throughout the world "fears and dangers of surprise attack." He proposed a joint agreement "for aerial photography of the other country," adding that the United States and the Soviet Union had the capacity to lead the world with mutually supervised reconnaissance overflights. The Soviets promptly rejected the proposal, saying that it was an obvious American attempt to "accumulate target information." Eisenhower later admitted, "We knew the Soviets wouldn't accept it, but we took a look and thought it was a good move."

Eisenhower bided his time and approved the development of reconnaissance satellites under the strictest security considerations. Virtually no one, even those in high national defense positions, knew of this effort. The WS 117L program was the prototype reconnaissance satellite effort of the United States. Built by Lockheed's Missile Systems Division in Sunnyvale, California, the project featured the development of a two-stage booster known as the "Agena" and a highly maneuverable satellite that took photographs with a wide array of natural, infrared, and other invisible light cameras. The first systems actually used film that was recovered after reentry from space but later ones employed the developing image transmission capabilities of television to send information electronically. The development contract was started in October 1956, with the intention of orbiting a working satellite by 1963. This early military space effort, while a closely guarded secret for years, was nonetheless critical to the later development of the overall U.S. space program.

SCIENTIFIC SATELLITES AND THE INTERNA-TIONAL GEOPHYSICAL YEAR. The civilian space effort of the Eisenhower administration began in 1952 when the International Council of Scientific Unions established a committee to arrange an International Polar Year, the third in a series of scientific activities designed to study geophysical phenomena in remote reaches of the planet. (*See Reading No. 2.*) The Council agreed that 1 July 1957 to 31 December 1958 would be the period of emphasis in polar research, in part because of a predicted expansion of solar activity; the previous polar years had taken place in 1882–1883 and 1932–1933. Late in 1952 this body expanded the scope of the scientific research effort to include studies that would be conducted using rockets with instrument packages in the upper atmosphere and changed the name to the International Geophysical Year (IGY) to reflect the larger scientific objectives. In October 1954 the Council, at a meeting in Rome, Italy, adopted another resolution calling for the launch of artificial satellites during the IGY to help map the Earth's surface, and both the Soviet Union and the United States accepted the challenge.

In response to these actions, on 26 May 1955 the National Security Council (NSC), the senior defense policy board in the United States, approved a plan to orbit a scientific satellite as part of the IGY effort. The NSC's endorsement was provisional:

the effort could not interfere with the development of ballistic missiles, must emphasize the peaceful purposes of the endeavor, and had to contribute to establishing the principle of "freedom of space" in international law. Eisenhower supported this effort and on 29 July publicly announced plans for launching "small unmanned, Earth circling satellites as part of the U.S. participation in the International Geophysical Year."

There followed a heady competition between the Naval Research Laboratory on the one hand and the Army's Redstone Arsenal on the other for government support to develop the IGY satellite. Project Vanguard, proposed by the Navy, was chosen on 9 September 1955 to carry the standard in launching a nonmilitary satellite for the IGY effort, over the Army's "Explorer" proposal. The decision was made largely because the Naval Research Laboratory candidate did not interfere with high-priority ballistic missile programs—it used the nonmilitary Viking rocket as its basis—while the Army's bidder was heavily involved in those activities and proposed adapting a ballistic missile launch vehicle. In addition, the Navy project seemed to have greater promise for scientific research because of a larger payload capacity. The Viking launch vehicle was also a proven system; it had first flown in the late 1940s while the Army's proposed rocket, the Redstone, had been launched for the first time in August 1953. Finally, the Naval Research Laboratory's proposal was more acceptable because it came from a scientific organization rather than a weapons developer, in the form of the Redstone Arsenal.

The architect of Project Vanguard, Naval Research Laboratory engineer Milton W. Rosen (1915–   ), called for the development and launching of six spacecraft for an estimated $20 million. The proposed launch vehicle combined the Viking first stage, an Aerobee sounding rocket second stage, and a new third stage with a 3.5 pound scientific satellite payload. Project Vanguard enjoyed exceptional publicity throughout the second half of 1955 and all of 1956, but the technological demands upon the program were too great and the funding levels too small to foster much success. Almost in desperation the laboratory launched the first Vanguard mission on 8 December 1956, a suborbital instrumentation test using a Viking rocket without the attendant second and third stages. Declared a successful flight, this mission nevertheless documented in graphic terms that the American IGY satellite effort was behind schedule; with the time left before the IGY there was little likelihood of orbiting a satellite unless resources approaching $100 million were applied to the program. Project Vanguard could not secure that kind of money from the parsimonious Department of Defense, which was supplying it from the secretary's emergency fund.

During the next several months the Eisenhower administration became increasingly concerned with the tendency for Project Vanguard to get bogged down. Eisenhower was especially concerned about the probability that the scientific instruments were slowing it down. About five months before the Soviet orbiting of Sputnik I, the president forcefully reminded his top advisors that "Such costly instrumentation had not been envisaged" and "stressed that the element of national prestige . . . depended on getting a satellite into its orbit, and not on the instrumentation of the scientific satellite." Eisenhower's perception of the budgetary growth of the Vanguard program, transforming it from the simple task of putting any type of satellite into orbit into a project to launch a satellite with "considerable instrumentation" reminded him of the worst type of technological inflation, as every scientist seemingly wanted to hang another piece of equipment on the vehicle.

Thus, less than a year before the launch of Sputnik I, the United States was involved in two modest space programs that were moving ahead slowly and staying within strict budgetary constraints. One was the highly visible scientific program, Vanguard, in honor of the IGY, and the other was a highly classified program to orbit a military reconnaissance satellite. They shared two attributes. They each were removed from the ballistic missile program underway in the DOD, but they shared in the fruits of its research and adapted some of its launch vehicles. They also were oriented toward satisfying the national goal of establishing "freedom of space" for all orbiting satellites. The IGY scientific effort could help establish the precedent of access to space, while a military satellite might excite other nations to press for closure. Because of this goal a military satellite, in which the Eisenhower administration was most interested, could not under any circumstances precede scientific satellites into orbit. The IGY satellite program, therefore, was a means of securing the larger goal of open access to space. Eisenhower was willing to place the military effort on simmer to ensure that scientific satellites led the way, hence his pressure on Vanguard. What became clear later was that he was not so concerned about orbiting the first satellite as he was about securing the precedent of free access to space for the United States.

SPUTNIK. The state of affairs of these two (civilian/military) space programs changed dramatically following the October-November 1957 launches of Sputniks I and II by the Soviet Union. On 4 October 1957 the Soviets launched Sputnik I from their rocket testing facility in the desert near Tyuratam in the Kazakh Republic. The world's first artificial satellite, Sputnik I weighed 183 pounds and traveled in an elliptical orbit that took it around the Earth every one and one half hours. (*See Reading No. 3.*) It carried a powerful radio beacon that beeped at regular intervals and could by means of telemetry verify exact locations on the earth's surface; some cold war hardliners suggested that this was a way for the Soviets to obtain targeting information for their ballistic missiles, but as yet there is little evidence to either confirm or deny this position.

Sputnik I twice passed within easy detection range of listening stations in the United States before anyone even knew of its existence. Then Moscow's official news agency, TASS, broke the story to the world. An IGY conference being held in Washington, D.C., learned the details of the launch the next day from the Soviet Union's chief delegate, Anatoli A. Blagonravov (1895–1975). The conference congratulated the Soviets for their scientific accomplishment. What was not said, but clearly thought by many Americans in both the scientific and political communities, however, was that the Soviet Union had staged a tremendous propaganda coup for the communist system, and that it could now legitimately claim leadership in a major technological field. The international image of the Soviet Union was greatly enhanced overnight.

While President Eisenhower and other leaders of his administration also congratulated the Soviets and tried to downplay the importance of the accomplishment, they misjudged the public

reaction to the event despite having been briefed on the psychological effect of such a Soviet accomplishment as early as 17
May 1955 at a National Security Council meeting. The launching of Sputnik I had a "Pearl Harbor" effect on American public
opinion. It was a shock, introducing the average citizen to space
affairs in a crisis setting. The event created an illusion of a
technological gap and provided the impetus for increased
spending for aerospace endeavors, technical and scientific educational programs, and the chartering of new federal agencies to
manage air and space research and development. Not only had
the Soviets been first in orbit, but Sputnik I weighed nearly 200
pounds, compared to the intended 3.5 pounds for the first
satellite to be launched in Project Vanguard. In the cold war
environment of the late 1950s, this disparity of capability
portended menacing implications. Even before the effects of
Sputnik I had worn off, the Soviet Union struck again. On 3
November 1957, less than a month later, it launched Sputnik II
carrying a dog, Laika. While the first satellite had weighed less
than 200 pounds, this spacecraft weighed 1,120 pounds and
stayed in orbit for almost 200 days.

THE AMERICAN RESPONSE. During the furor surrounding these events many people accused the Eisenhower
administration of neglecting the American space program. The
Sputnik crisis reinforced for many people the popular conception that Eisenhower was a smiling incompetent; it was another
instance of a "do-nothing," golf-playing president mismanag-
ing events. G. Mennen Williams (1911–   ), the Democratic
governor of Michigan, even released a poem about it:

Oh little Sputnik, flying high
With made-in-Moscow beep,
You tell the world it's a Commie sky
and Uncle Sam's asleep.

You say on fairway and on rough
The Kremlin knows it all,
We hope our golfer knows enough
To get us on the ball.

More seriously, Senator Lyndon B. Johnson (1908–1973),
Democrat-Texas, opened hearings by a subcommittee of the

Senate Armed Services Committee on 25 November to review the whole spectrum of American defense and space programs in the wake of the Sputnik crisis. This group found serious under-funding and incomprehensible organization for the conduct of space activities. It blamed the president and the Republican Party. One of Johnson's aides, George E. Reedy (1917–   ), summarized the feelings of many Americans: "The simple fact is that we can no longer consider the Russians to be behind us in technology. It took them four years to catch up to our atomic bomb and nine months to catch up to our hydrogen bomb. Now we are trying to catch up to their satellite." The administration now had to move quickly to restore confidence at home and prestige aboard. The response was typical of earlier crises within the United States; politicians locked arms and appropri-ated money to tackle the perceived problem. In this effort both the civilian and military space efforts benefited, one openly and the other still in secrecy.

As the first tangible effort to counter the apparent Soviet leadership in space technology, the White House announced that the United States would test launch a Project Vanguard booster on 6 December 1957. The media was invited to witness the launch in the hope that it could help restore public confidence, but it was a disaster of the first order. During the ignition sequence the rocket rose about three feet above the platform, shook briefly, and disintegrated in flames. The next test was little better. On 5 February 1958 the Vanguard launch vehicle reached an altitude of four miles and then broke apart. Public perceptions of American technological capabilities were ex-tremely low after these two failures.

In the crisis situation near the end of 1957, the Department of Defense approved an additional $13 million for an Army effort, featuring Wernher von Braun and his German rocket team laboring with their rocket to launch two satellites. The Army's Explorer project had been shelved earlier in favor of concentrat-ing on Vanguard, but drastic times called for drastic measures and suddenly the atmosphere in Washington had changed. The Army was told to orbit the first satellite by 30 January 1958, only four months after the first Sputnik. Von Braun and his team went to work on a crash program with a modified Jupiter C ballistic missile. They constructed a four-stage launch vehicle, renamed after some modifications Juno 1, and a satellite payload called Explorer.

The first launch took place on 31 January 1958, a day later than called for by the Eisenhower administration, placing Explorer 1 in orbit. On this satellite was an experiment by James A. Van Allen (1917–   ), a physicist at the University of Iowa, documenting the existence of radiation zones encircling the earth. Shaped by the earth's magnetic field, what came to be called the Van Allen radiation belt partially dictates the electrical charges in the atmosphere and the solar radiation that reaches earth. A long series of Explorer satellites, each with a different design and purpose, yielded valuable scientific information during succeeding years.

Project Vanguard also received additional funding to accelerate its activity, and Vanguard 1 was finally orbited on 17 March 1958. Vanguard 1 and its successor satellites in 1959 conducted geodetic studies of the earth, explored further the Van Allen radiation belt, and discovered the earth's slightly "pear" shape. Both the Vanguard and the Explorer series of satellites were enormously important in collecting scientific information about the universe. They were also partially successful in salving the nation's wounds at not being *first* in orbiting a satellite.

EISENHOWER AND THE SECURING OF FREEDOM OF SPACE. While these activities were taking place, the Eisenhower administration was working behind the scenes to achieve permanent free access to space and to avoid international overflight issues common to aviation. He was concerned, as already mentioned, that if the United States was the first nation to orbit a satellite the Soviet Union could invoke territorial rights in space. Soviet Sputniks I and II, however, had overflown international boundaries without provoking a single diplomatic protest. On 8 October 1957, for example, Deputy Secretary of Defense Donald Quarles told the president: "The Russians have . . . done us a good turn, unintentionally, in establishing the concept of freedom of international space." Eisenhower immediately grasped this as a means of pressing ahead with the launching of a reconnaissance satellite. The precedent held for Explorer 1 and Vanguard 1, and by the end of 1958 the tenuous principle of "freedom of space" had been established. By allowing the Soviet Union to lead in this area, the Russian space program had established the U.S.-backed precedent for free access.

During 1958 Eisenhower's National Security Council issued directives that established critical goals for the U.S. space program, none of which were civilian. It affirmed the free-access-to-space position already established in precedent and declared that space would not be used for warlike purposes. At the same time it asserted that reconnaissance satellites and other military support activities that could be aided by satellites, such as communications and weather, were peaceful activities since they assisted in strategic deterrence and therefore averted war. This was a critical space policy as it provided for open use of space and fashioned a virtual "inspection system" to forewarn of surprise attack through the use of reconnaissance satellites. With these goals affirmed and tacitly agreed to in the international arena, the United States proceeded at an accelerated pace with the development of military reconnaissance satellites. It did so with the same degree of secrecy the WS 117L program had enjoyed to that point. The United States also moved forward on the civilian side of space endeavors, but in this case with much fanfare and public support.

CHAPTER 3

CREATING A UNIFIED SPACE PROGRAM

LAUNCHING NASA. Less than two weeks after the orbiting of Sputnik I in October 1957 Senator Lyndon B. Johnson made a speech calling for a congressional review of the American space effort and for all Americans "to work together as we step into a new age of history." It was critical, Johnson believed, to make space exploration a concerted effort both for technology development and for the national prestige it would engender. In an 11 April 1958 memorandum Johnson's congressional review found that "The reason the United States fell behind Russia in satellite development in the first place is because we neglected the relation between scientific achievement and international relations." Johnson early committed to a course that led to the creation of an independent organization that encompassed both areas of concern.

Two major issues immediately emerged in Johnson's investigation. First, his Senate review ascertained the status of existing space-related activities and found them wanting. The Eisenhower administration, however, took steps to assure that progress sped up, and the result was the launch of Explorer 1 on 31 January 1958. Second, the congressional inquiry also assessed the nature, scope, and organization of the nation's long-term efforts in space. The findings were bleak on this score, and as a direct result on 6 February 1958 the Senate voted to create a Special Committee on Space and Aeronautics whose charter was to frame legislation for a permanent space agency. The House of Representatives soon followed suit. With Congress leading the way, and fueled by the crisis atmosphere in Washington following the Sputnik episode, it was obvious that some government organization to direct American space efforts would emerge before the end of the year.

The principal questions concerning any new space agency in the first half of 1958 revolved around whether it should be civilian or military in orientation and organization, whether it should be an existing or newly created entity, and how aggressive it should be in exploring space. Congress first considered assigning the space mission to several existing agencies. The Department of Defense (DOD) was an early and especially attractive contender because of its longstanding work with both military rockets and scientific Vanguard and Explorer missions. The Atomic Energy Commission (AEC), which had experience with nuclear missile warheads, also had support in Congress for this broader mission. When Senator Clinton Anderson (1895–1975) (Democrat-New Mexico) introduced legislation in January 1958 to give primacy in the space program to the AEC, Eisenhower and his supporters came forward to oppose it because of the agency's military involvements. To do so effectively, however, they had to devise an alternative plan that made sense.

On 4 February 1958 the president asked his new science advisor, James R. Killian (1904–1988), to convene the President's Science Advisory Committee (PSAC), established in the wake of Sputnik, to come up with a plan. (*See Readings No. 4 and 5.*) A month later Killian came forward with a proposal that placed all nonmilitary efforts relative to space exploration under a strengthened and renamed National Advisory Committee for Aeronautics (NACA). Established in 1915 to foster aviation progress in the United States, the NACA had long been a small, loosely organized, and elitist organization known for both its technological competence and its apolitical culture. It had also been moving into space-related areas of research and engineering during the 1950s; for instance, its engineers developed the blunt-shaped reentry concept used in the recovery of space capsules. Scientists were also involved in testing the aerodynamics of rocket models by firing them out over the Atlantic Ocean from a launch site at Wallops Island, Virginia. While totally a civilian agency the NACA also enjoyed a close working relationship with the military services, helping to solve research problems associated with aeronautics and also finding application for them in the civilian sector. Its civilian character; its recognized excellence in technical activities; and its quiet, research-focused image all made it an attractive choice. It could fill the requirements of the job without exacerbating cold war tensions with the Soviet Union.

President Eisenhower accepted the PSAC's recommendations and had members of his administration draft legislation to expand the NACA into a new National Aeronautics and Space Administration (NASA). It set forth a broad mission for the agency to "plan, direct, and conduct aeronautical and space activities"; to involve the nation's scientific community in these activities; and to disseminate widely information about these activities. An administrator appointed by the president was to head NASA, a departure from the committee approach that had characterized the NACA. The president wanted a single-headed agency who would report directly to him. As a compromise the proposed structure also called for the creation of an advisory board of seventeen members who had no administrative responsibility.

Lyndon Johnson was not entirely pleased with the Eisenhower administration's proposal. He argued that its chief weakness was the lack of a central policy-making body that had the authority to mediate conflicts between the civilian and military space efforts and to ensure a cohesive effort. With knowledge of the interservice rivalries that had plagued the military rocket programs, Johnson warned that NASA did not have the "authority over the entire space program so that it can be handled with foresight rather than on a trouble-shooting basis." A Senate special committee memo of 11 April 1958 stated the problem this way: "The bill provides for cafeterias for NASA but not for overall coordination and control within our own government so that our relations with the Soviet Union on satellites will be handled with foresight." Johnson inserted into the proposed legislation language that created what was eventually called the National Aeronautics and Space Council of no more than nine members charged with working out "the aeronautic and astronautic policies, programs and projects of the United States." Required seats on the council included the NASA administrator, the secretaries of State and Defense, and the head of the AEC.

Eisenhower was opposed to the Space Council at first. He feared that it would become a powerful new organization on the order of the National Security Council that would require too much of his attention. Eisenhower and Johnson squared off over this issue in the legislative arena and the initial result was a predictable standstill. Eisenhower tried to explain that he wanted a purely advisory body at most; Johnson wanted one that would be able to cut through bureaucratic layers and establish direction for the space program. Eisenhower finally decided to meet Johnson to discuss the issue on Sunday, 7 July 1958. Their short meeting ended in compromise. When Eisenhower suggested that the Space Council had the potential to place too many demands on him, Johnson suggested that the president chair it himself, thereby setting its agenda and channeling its efforts along courses he approved. Eisenhower agreed to this stipulation because he wanted, as he later told Killian, "to see the bill move ahead." The stalemate ended almost immediately, and the National Aeronautics and Space Act of 1958 was passed by Congress. (*See Reading No. 6.*) Eisenhower signed it into law on 29 July 1958 and the new organization started functioning on 1 October.

BUILDING A SPACE AGENCY. The 170 employees of the new space organization gathered in the courtyard of the Dolly Madison House near the White House in downtown Washington on 1 October 1958 to listen to the newly appointed NASA Administrator, T. Keith Glennan (1905–   ), announce the bold prospects being considered for space exploration. Glennan, fresh from the presidency of the Case Institute of Technology in Cleveland, Ohio, presided over a NASA that had absorbed the NACA intact, and the old NACA of about 8,000 employees and an annual budget of $100 million made up the core of the new NASA. It consisted of a small headquarters staff in Washington that directed operations. It also had three major research laboratories—the Langley Aeronautical Laboratory established in 1918, the Ames Aeronautical Laboratory activated near San Francisco in 1940, and the Lewis Flight Propulsion Laboratory built at Cleveland, Ohio, in 1941—and two small test facilities, one for high-speed flight research at Muroc Dry Lake in the high desert of California and one for sounding rockets at Wallops Island, Virginia. The scientists and engineers that came into NASA from the NACA brought a strong sense of technical competence, a commitment to collegial in-house research conducive to engineering innovation, and a definite apolitical perspective.

Within a short time after NASA's formal establishment, several organizations involved in space exploration projects from other federal organizations were incorporated into NASA. One of the important ingredients was the 150 personnel, mission, and resources associated with Project Vanguard at the Naval Research Laboratory located along the Potomac River just outside of Washington. Officially becoming a part of NASA on 16 November 1958, this group remained under the operational control of the Navy until 1960 when it was transferred en masse from Navy facilities to a newly established NASA installation, the Goddard Space Flight Center, located outside Washington in suburban Maryland. Those who had been associated with the Naval Research Laboratory brought a similar level of scientific competence and emphasis on in-house research and technical mastery that had been the hallmark of the NACA elements.

In addition to the Project Vanguard personnel and resources, NASA quickly gained several disparate satellite programs, two lunar probes, and the important research effort to develop a million-pound-thrust, single-chamber rocket engine from the Air Force and the DOD's Advanced Research Projects Agency. In December 1958 NASA also acquired control of the Jet Propulsion Laboratory, a contractor facility operated by the California Institute of Technology (Caltech) in Pasadena, California. Coming from the Army, this oddly-named institution had been specializing in the development of weaponry since World War II.

During this period of rapid expansion, Glennan also asked for the transfer to NASA of part of the Army Ballistic Missile Agency (ABMA), part of the Redstone Arsenal, located at Huntsville, Alabama, and presided over by one of the nation's foremost space advocates, German postwar immigrant Wernher von Braun. The Army had dug in its heels, however, and refused to give up the jewel in the crown of its space vision. Von Braun's German rocket team, as it was called, numbered only about 100 people, but it was firmly in control of the 4,500 person installation at Huntsville and it was the Army's centerpiece in an interservice struggle for the space mission. The Army pinned high hopes on ABMA's most important project, the development of a rocket that could deliver 1.5 million pounds of thrust in the first stage, eventually named the Saturn. The Saturn would without question establish the Army's leadership in the development of space technology.

The Army resisted NASA's overtures for 18 months. During the summer of 1959, however, congressional criticism forced the DOD to reevaluate the Army's Saturn program. The Army, within its assigned military mission, had no business developing this super space booster. If there was a military use, which was problematic, it clearly rested within the Air Force's mission rather than the Army's. NASA exploited this criticism and suggested that it definitely had a use for the Saturn launch vehicle, so a transfer of all personnel and resources associated with the project to the space agency would be appropriate. Additionally, Glennan also argued that transfer of the Saturn to NASA would avoid interservice rivalries, which were always tense, since the DOD would not have to choose between the Army and the Air Force. Accordingly, on 1 July 1960 the ABMA's shift from the Army to NASA with all personnel and resources intact was completed. It was renamed the George C. Marshall Space Flight Center in honor of the American army officer and secretary of defense who had helped win World War II and then rebuild Europe. This rocket team brought to NASA a strong sense of technical competence, a keen commitment to the goal as defined by von Braun, and an especially hardy group identity.

By mid-1960 NASA had gained primacy in the federal government for the execution of all space activities except reconnaissance satellites, ballistic missiles, and a few other projects, most of which were still in the study stage, that the DOD still controlled. These military missions were still considerable, however, and accounted for over half of the federal budget spent on the space effort at the time. The clear mandate from the Eisenhower administration, it should be emphasized, was that NASA's space efforts would be both nonmilitary in character and highly visible to the public. This would serve two distinct but necessary purposes. First, NASA's projects were clearly cold war propaganda weapons that national leaders wanted to use to sway world opinion about the relative merits of democracy versus the communism of the Soviet Union. The rivalry was not friendly, and the stakes were potentially quite high, but it at least this competition had the virtue of not being military in disposition. It was not, after all, an arms race and the likelihood of any aspect of it leading to war and potentially to nuclear destruction was slim. Second, NASA's civilian effort served as an excellent smoke-screen for the DOD's military space activities, especially for reconnaissance missions. NASA's civilian mission, therefore, dovetailed nicely into cold war rivalries and priorities in national defense.

GOOD NEWS AND BAD NEWS FOR SATELLITES. Even as NASA was getting organized, it was expected to carry out a rather broad mandate of civilian space exploration. It did so by conducting several satellite programs aimed either at gathering scientific data about the solar system or providing weather, communications, or other types of services to the nation. Many of these activities were conducted on a trial and error basis during those early years. Reliability was always a problem. Many times the rockets boosting satellites into orbit did not perform properly, and even if the satellite was successfully launched it might not function properly. For example, in 1958 only 5 of 17 launch attempts were successful, and the next year 10 of 21 were effective. Improvement was steady thereafter. By 1963 60 of 71 launches were successful. Between 1958 and 1965 the United States went from a 71 percent launch failure rate to a 91 percent success rate. Slow, steady improvement in the reliability of launch vehicles was largely responsible for this change.

NASA continued the programs begun by the Army and Navy, Explorer and Vanguard, and began some initiatives of its own during those early years. The Army launched five Explorer satellites, each with a different type of scientific payload aboard, before the program transferred to NASA. Under NASA's aegis an additional 11 Explorer satellites were launched through the end of 1961, all but three of which orbited successfully and returned scientific data. The Vanguard program was more star-crossed. After Vanguard 1's launch on 17 March 1958, Vanguard 2 was launched under NASA auspices on 17 February 1959; it did not achieve the desired orbit and collected data for only 18 days. Two unsuccessful launch attempts in April and June 1959 raised further questions about the program, but NASA stayed with it. Vanguard 3 was finally placed in orbit in September 1959, contributing additional data on the Van Allen radiation belts before termination of the program.

Two important initiatives also sent scientific satellites to other bodies in the solar system, another arena of competition for technological firsts between the United States and the Soviet Union. The first of these was Pioneer, the maiden voyage of which began on 11 October 1958. Unfortunately, the second and third stages did not separate properly and the probe did not achieve a Lunar trajectory. Four more Pioneer spacecraft were launched through 1960, but only Pioneer 5, launched on 1 March 1960, gathered much scientific information. It transmitted the first solar flare data and established a communications distance record of 36.2 million kilometers. No additional Pioneers were launched until 1965. On 12 September 1959 the Soviet space probe, Luna 2, sent back the first clear images of the Lunar surface and then impacted on it. An American lunar probe was a direct reaction to this success. Initiated by NASA, Project Ranger was to photograph and map the Moon's surface. This would not only provide a great amount of scientific data about the Moon, but would also pave the way for human exploration. NASA launched Ranger 1 on 23 August 1961, but it failed to achieve its planned orbit. A total of nine Rangers were launched through 1965, but only the last three were successful in sending data and impacting the Moon. Once again, the United States seemingly lagged behind the Soviet Union in the cold war battlefield of technological competition in space.

One area where the United States made significant strides during the early years of space exploration, however, was in application satellites. In 1945 science fiction writer and futurist Arthur C. Clarke had posited that three satellites placed in geosynchronous (stationary) orbit 22,240 miles above the equator, could be used to bounce radio waves around the globe. The idea thrilled many scientists and with the dawning of the space age NASA began an effort to make it a reality. The first attempt was a test program called Echo, which called for the orbiting of a 100-foot inflatable satellite covered with reflective material that NASA could bounce a radio beam off. Difficulties abounded in trying to launch an inflatable, passive satellite, but tests were successful on 12 August 1960. At the same time active-repeater communications satellites were being developed, the first of which was the Bell Telephone Laboratories Telstar project. Beginning in 1962, several generations of Telstars, as well as other types of communications satellites in Earth orbit, helped to make Clarke's idea of real-time global communications a reality by the mid-1960s.

Seeing the enormous commercial potential of space-based communications, the U.S. Congress passed the Communications Satellite Act of 1962, creating Communications Satellite Corporation (Comsat) with ownership divided fifty/fifty between the general public and the telecommunications corporations to manage global satellite communications for the United States. Near the same time U.S. leaders recognized the possibility of competition and participated in the establishment of the International Telecommunication Satellite Consortium (INTELSAT), with Comsat as manager, to provide an international communication satellite system. Founded by nineteen nations, with eventual membership of well over a hundred, it was initially very much an American organization, with the United States controlling 61 percent of the voting authority and all the technology. It oversaw the development of INTELSAT 1 in 1965, the first of the global communications satellite network. With this satellite system in orbit the world became a far different place. Within a few years telephone circuits increased from five hundred to thousands and live television coverage of events anywhere in the world became commonplace.

Equally successful was NASA's work with meteorological satellites, providing weather data from space. Project Tiros had been inherited by NASA from the DOD during the administrative consolidation between 1958 and 1960. The space agency launched Tiros 1 on 1 April 1960 and it provided valuable images of weather fronts, storms, and other atmospheric occurrences. This satellite led to a long series of weather satellites that quickly became standard weather forecasting tools in the United States and throughout the world.

THE HUMAN SPACEFLIGHT IMPERATIVE. Even as diverse organizational elements were being incorporated into NASA and the satellite program was experiencing both successes and failures, the space agency was beginning to develop plans for the orbiting of an American around the Earth. This was an especially attractive idea since the Soviet Union had already won most of the battles for prestige associated with satellites. Human spaceflight, therefore, became the next arena for competition between the United States and the Soviet Union in the cold war. All of the uniformed services got into the act and even before the creation of NASA they had developed plans for sending an American into space.

The Air Force project, with the unlikely name of Man in Space Soonest (MISS), advocated a four-part plan to land humans on the Moon by the end of 1965 using existing military boosters at the bargain basement price of $1.5 billion. Even more far-fetched was the Navy's proposal to orbit a novel spacecraft, a cylinder with spherical ends that would telescope into a delta-wing, inflated glider with a rigid nose section. The Manned Earth Reconnaissance (MER) program was an innovative idea that had little chance of success because of its emphasis on new hardware and entirely unexamined techniques, and it was quickly derailed. The Army's entry into the human spaceflight sweepstakes was devised at ABMA by von Braun and his rocket team. Much simpler and less ambitious than the Air Force and Navy plans, Project Adam called for the use of a modified Redstone booster to launch a pilot in a sealed capsule along a steep ballistic, suborbital trajectory. The capsule would reach an altitude of about 150 miles before splashing down by parachute in the Atlantic missile range east of Cape Canaveral, Florida, where von Braun had established ABMA's launch facilities. In spite of the seriousness with which the Army put forth this plan, many agreed with Dr. Hugh L. Dryden (1898–1965), the NACA's director of research, that it had "about the same technical value as the circus stunt of shooting the young lady from the gun. . . ."

The coup de grâce to each of these programs came with the actual creation of NASA and its presidentially approved assignment of putting an American in space. A few NACA engineers under the leadership of Dr. Robert R. Gilruth (1913–) at Langley Aeronautical Laboratory had initiated work on the possibility of a piloted spacecraft in the spring of 1958 and just days after NASA was officially activated they proposed Project

Mercury to Administrator Glennan. They advocated a three-phased project: (1) Redstone missiles would be used to send humans on suborbital ballistic flights very similar to what Dryden had termed a "circus stunt"; (2) Jupiter rockets would be used for longer suborbital trajectories, a phase later scrapped as unnecessary; and (3) Atlas boosters would launch an occupied capsule into orbit. In all, the project called for six piloted flights. Emphasizing the use of existing technology, relative simplicity, and a progressive and logical testing program the plan realistically aimed toward the objective of putting a human in orbit within two years. The NASA chief gave his blessing, a decision soon approved by the National Aeronautics and Space Council. Glennan established a Space Task Group under Gilruth to supervise the Mercury program. This group in 1962 moved to a new research installation, the Manned Spacecraft Center (renamed the Lyndon B. Johnson Space Center in 1973), near Houston, Texas. The decision to locate the center in Houston, it should be mentioned, apparently resulted from the influence of House Speaker Albert Thomas, who represented that district in Congress.

During the months that followed approval of Project Mercury, the Space Task Group energetically pursued the development of the hardware and support structure to handle the program. Dr. Maxime A. Faget (1921–   ) was the chief designer of the Mercury spacecraft, a compact vehicle capable of sustaining a single person in orbit for about 24 hours. Late in 1959 the McDonnell Aircraft Corporation was chosen as the prime manufacturer of the Mercury. The Mercury spacecraft was about 11 feet long and 6 feet wide at the base, conical in shape. Designed to orbit with the astronaut seated facing backward, it would be slowed by a retrorocket pack to allow gravity to bring it back to Earth. The spacecraft came down to an ocean landing braked by parachutes. The astronaut had very little room for movement, being placed in an individually fitted contour seat for the duration of the flight. Engineers began integrating the boosters and the spacecraft into a unit that would operate reliably together during 1960. This effort was aided by the transfer of ABMA to NASA and the easier ability to tap the expertise of the builders of the Redstone rocket. Additionally, NASA expanded the infrastructure supporting spaceflight operations by establishing ground tracking stations around the globe, a mission control center, a complex communications system, and by expanding the launch complex at Cape Canaveral.

The first Mercury test flight took place on 21 August 1959 when a capsule carrying two rhesus monkeys was launched atop a cluster of Little Joe solid-fuel rockets. Other tests using both Redstone and Atlas boosters and carrying both chimpanzees and astronaut dummies soon followed. For instance, on 31 January 1961 the chimpanzee Ham flew 157 miles into space in a 16-minute, 39 second flight in a Mercury/Redstone combination and was successfully recovered.

Concurrent with this effort NASA selected and trained the Mercury astronaut corps. Contrary to a NASA priority that these six astronauts be civilians, President Eisenhower directed that they come from the armed services' test pilot force. A grueling selection process began in January 1959, and early in April, Gilruth and company narrowed the list to seven candidates. Unable to cut the last candidate, NASA decided to appoint seven rather than six men as astronauts. Glennan and Washington politicos publicly unveiled the astronauts in a circus-like press conference on 9 April 1958. The seven men—from the Marine Corps, Lt. Col. John H. Glenn, Jr. (1921–   ); from the Navy, Lt. Cdr. Walter M. Schirra, Jr. (1923–   ), Lt. Cdr. Alan B. Shepard, Jr. (1923–   ), and Lt. M. Scott Carpenter (1925–   ); and from the Air Force, Capt. L. Gordon Cooper (1927–   ), Capt. Virgil I. "Gus" Grissom (1926–1967), and Capt. Donald K. Slayton (1924–1993)—became heroes in the eyes of the American public almost immediately, due in part to a deal they made with *Life* magazine for exclusive rights to their stories.

The astronauts essentially became the personification of NASA to most Americans during the Mercury project, creating internal jealousies and turf battles over who controlled the spaceflight program. With their celebrity status they exercised important influences over the direction of the program; for instance they forced changes in the design of the Mercury capsule to permit greater pilot control. Despite the periodic irritations among astronauts, engineers, and administrators, it should be mentioned that disagreements were the exception rather than the rule and all parties spent most of their time working together to further the project's operational readiness.

Even as these efforts were underway, the realities of the cold war emerged to shake NASA out of its "ivory tower" of deliberate scientific and technological development and transform it into a powerful vehicle for competing with the Soviet Union. Heretofore, in line with Eisenhower's priorities, the agency had avoided racing the Soviets. In a remarkable statement, T. Keith Glennan, the NASA administrator, confided in his diary on 1 January 1960 that he would not conduct the activities of the U.S. space program in response to the Soviet Union. "We are not going to attempt to compete with the Russians on a shot-for-shot basis in attempts to achieve space spectaculars," he wrote, adding, "Our strategy must be to develop a program on our own terms which is designed to allow us to progress sensibly toward the goal of ultimate leadership in this competition." While a rational position, this did not take into consideration the harshness of the cold war environment of the early 1960s and the seemingly life-and-death struggle between the two superpowers. That competition ultimately dictated the activities of NASA for the balance of the decade, especially in relation to the United States no-holds-barred sprint to the Moon known as Project Apollo.

CHAPTER 4

THE RISE OF SPACE SCIENCE
AND TECHNOLOGY

To some Americans, from the 1960s to the present, space has represented prestige and the American image on the world stage. To others it has signified the quest for national security. To still other Americans, space is or should be, about gaining greater knowledge of the universe. It represents, for them, pure science and the exploration of the unknown. Even so, the history of space science and technology is one of the largely neglected aspects in the history of the space program. This chapter explores some of the early space satellite efforts aimed at multiplying scientific knowledge, and the development of technology applicable to the space program as manifested in the evolution of rockets. It also investigates some of the contentions between groups involved in the field and NASA's role in developing scientific disciplines during the decade of the 1960s.

LAUNCH VEHICLE DEVELOPMENT. The explosive growth of technology on a broad front may well be one of the most significant attributes of the United States in the twentieth century. During the 1960s several strains of technological research and development (R&D) at NASA and elsewhere converged to enhance the development of rockets used for space science and other applications. Several military missile programs; smaller research efforts by the old National Advisory Committee for Aeronautics (NACA), the Jet Propulsion Laboratory, and the Naval Research Laboratory; and a few private and university rocket research programs set the stage for the development of a family of reliable launch vehicles in the 1960s and pointed the direction toward the development of the crowning vehicle produced in the decade, the Saturn rocket, to be discussed later. Less ambitious steps were important, however, and they point up well the application of technology in one area.

In virtually every instance rockets used in the space program during the decade resulted from the adoption of a basic system built on components that had been tested earlier and mated together into a new booster. For instance, the Scout booster began in 1957 as an attempt by the NACA to build a solid-fuel rocket that could launch a small scientific payload into orbit. To achieve this end, researchers investigated various solid-rocket configurations and finally decided to combine a Jupiter Senior (100,000 pounds of thrust), built by the Aerojet Corporation, with a second stage composed of a Sergeant missile and two new upper stages descended from the research effort that produced the Vanguard. The Scout's four-stage booster could place a 330 pound satellite into orbit, and it quickly became a workhorse in orbiting scientific payloads during the era. It was first launched on 1 July 1960, and despite some early deficiencies, by the end of 1968 had achieved an 85 percent launch success rate. Later versions of the Scout were still being used productively in the early 1990s.

While the Scout became an outstanding small booster, more than any other factor the lack of a reliable heavy-lift launch vehicle hampered the exploration of space in the decade of the 1960s. The Soviet Union, as demonstrated in its early space program successes, was somewhat ahead of the U.S. in rocket technology in 1960. Resolving that discrepancy and advancing scientific goals prompted NASA engineers to modify existing boosters for greater lift. As one example, NASA adopted the liquid-hydrogen fueled Thor, originally developed by the U.S. Air Force as a ballistic missile, as a first stage booster and added an Agena upper stage, another Air Force missile, to create a very successful combination that could launch a 2,200 pound scientific payload to orbit. It was first used by the U.S. Air Force on 28 February 1959 to launch Discoverer 1, and later versions placed several scientific, weather, reconnaissance, and communications satellites in orbit for NASA.

The most successful family of launch vehicles to be developed in the 1960s, however, was the Delta. A modification of the Thor booster in its initial stages, the Delta had two upper stages derived from the Vanguard booster. Together these could orbit medium-sized payloads. First ready for use by 1960, the Delta system had achieved 150 launches through 1979 with a success rate of 92 percent. The vehicle greatly benefitted in 1972 from the addition of solid-fuel "strap-on" Scout boosters that enabled it to launch 4,300 pounds of payload into orbit. Concomitant with the development of that launch system, the United States developed another family of boosters, the Titan, which was also enormously successful. Using another ICBM launch vehicle as a basis, this system began to emerge in the mid-1960s as the booster of choice for all heavy-payload space exploration save the human spaceflight program.

Because of this concerted effort, by the middle part of the 1960s the United States had developed several launch vehicle families that could deliver payloads when and where they were needed. The very success of these efforts, however, meant that additional technological innovation was stunted since the phrase, "if it ain't broke, don't fix it," came to rule NASA and others involved in launch vehicle development during the austere 1970s. The present fleet of launch vehicles used by the military and civil space efforts, even such presumably modern systems as that developed for NASA's Space Shuttle, were predicated on 1960s technological breakthroughs. Meantime, other nations moved to catch up and surpass the United States in space booster technology, and such vehicles as the European Space Agency's Ariane family and the Japanese H-2 were the result. The tortoise and hare competition has continued to the present in launch vehicle development as the United States decided to catch up once again with foreign technology by announcing in 1991 the development of a National Launch System.

SEIZING THE INITIATIVE. The modern components of U.S. space science emerged rapidly in the wake of Sputnik I in late 1957. The National Aeronautics and Space Administration was explicitly charged in its congressional act with "the expansion of human knowledge of phenomena in the atmosphere and space." The legislation also directed the NASA administrator to organize for the development of space science programs in coordination with scientists at universities and other institutions. In fulfillment of this mandate, in the fall of 1958 NASA established at its headquarters a space science office and installed as its head the respected mathematician and atmospheric scientist, Homer E. Newell (1915–1983), brought over from the

Naval Research Laboratory. During the next several years the place of Newell's bailiwick in the NASA organization—as well as its size, scope, and method of operations—was hammered out both within the institution and in the outside scientific community. When James E. Webb created the Office of Space Sciences with Newell at the head not long after taking over as NASA administrator in 1961, he gave the effort the kind of visibility that its clients thought was appropriate.

Newell established a broad-based method for choosing space science experiments and allocating support—be it financial or otherwise—for them. He solicited proposals for projects from research facilities, educational institutions, other government organizations, the National Academy of Science's Space Science Board, and industry. These would then be considered, along with proposals from NASA's scientists, for adoption and funding by NASA. Participants in any given space science research project usually included representatives from each of the major constituencies involved in the solicitation process. Most important, each chosen project was to be executed under the direction of a NASA program scientist who was, according to a memorandum on the approach issued by Newell in 1960, "generally responsible for the overall coordination of the activities of the various participants. . . . and will have responsibility and authority for resolution of any disagreements between and among various participants." (*See Reading No. 7.*) This approach toward conducting space science placed NASA as an institution in the driver's seat of virtually all space science in the United States. Certainly, all scientific experiments that required the flight of a payload would have to adhere to its policies since NASA had a virtual monopoly on launch capability for scientific pursuits. Scientists working in space science activities not dependent on spacecraft—ground-based astronomy, etc.—also found it increasingly important to adhere to NASA's scientific priorities in order to obtain the much needed funding that was available from the space agency.

This was a bitter pill for scientists to swallow, and during the first four years after the creation of NASA, the scientific community fought a series of internecine battles over who would control the resources and therefore the direction of space science. Scientists resisted reporting to government officials on the conduct of their projects, and, more important, they objected to the loss of autonomy the NASA research program necessitated. The Space Science Board of the National Academy of Sciences became the vehicle of choice for outside (mostly university) scientists to combat NASA's control over scientific projects. This pressure from the university community first met with resistance and then compromise between the parties during the following years. At no time, however, did the agency delegate the authority for planning and executing space science to another body. In spite of some rocky disturbances early in NASA's history, therefore, the relationships between members of the scientific community was by the early 1960s relatively stable and collegial, cobbling together a NASA/university/industry/research installation partnership to execute a broad range of scientific activities in the 1960s. While enjoying an uneasy relationship, NASA had seized the initiative in space science.

APPLYING SPACE SCIENCE TO THE STUDY OF PLANET EARTH. As 1962 began, Homer Newell had control of the institutions, resources, and facilities necessary to execute a wide-ranging program in space science. The availability of satellites in orbit around the earth during the 1960s made it possible to expand scientific understanding of this home planet's size, geology, organic and inorganic construction, and physical properties. Prior to this time Earth scientists of all types had developed their ideas solely on the basis of in-depth field work but had not been able to observe from afar. That changed with the advent of satellites as they could now review from a macro level areas difficult to observe from the ground or the air. Instead of now being mired in the trees, so to speak, scientists concerned with all manner of Earth features could employ remote sensing to observe the forest. This capability proved revolutionary.

Using satellites, scientists were able in the 1960s to undertake pathbreaking geodetic research. They measured the Earth as never had been possible before, increasing by a factor of 10 knowledge of its size and shape, and by a factor of 100 an understanding of its gravitational field. By 1970 a worldwide geodetic net had also been established, allowing common reference points to be established anywhere on the globe with an accuracy of 15 meters. An important outgrowth of this satellite research, although other scientific approaches also contributed, was the theory of plate tectonics advanced in the latter 1960s to explain the dynamics of the Earth's outer shell. The theory posited that the Earth's surface, the lithosphere, consists of about a dozen large plates and several small ones that move relative to each other and interact at their boundaries. This theory, which now has broad acceptance in the scientific community, goes far toward explaining seismic and volcanic activity as well the origins and evolution of mountains and other geographical features. The aggregate of this research has been significant; for example, in a single decade more information on the Earth's physical properties had been gathered and digested than in the previous 200 years.

Atmospheric, ionospheric, and meteorological science also benefitted greatly from the opportunity to study the Earth by satellites. Building steadily on the research base of earlier years, especially the rocketry experiments conducted before the creation of NASA, by the 1970s, Homer Newell wrote in 1980, "all known major problems of the high atmosphere and ionosphere had a satisfactory explanation based on sound observational data." In meteorology the advance was not as dramatic, but it was persistent. The perspective afforded by satellite imaging was a great boon to scientists who were then able to locate distribution and kinds of cloud formations, find and measure weather disturbances, and track movements and patterns. It provided new levels of precision to the evaluation of pressure fronts and air masses that are so critical in weather forecasting. Likewise, meteorological research beyond weather forecasting took on new life as climatological research contributed significant insights to an understanding of Earth.

One of the genuine products of the space age and its new research vehicles involved the development of magnetospheric physics. James A. Van Allen's 1958 Explorer 1 experiment had established the existence of radiation belts around the Earth, representing the opening of a broad research field. The discovery of a terrestrial magnetosphere, as well as other important breakthroughs—especially using the Explorer series of spacecraft, 34 of which were launched in the 1960s—explained much about the relationship of the Sun to the Earth, the interplanetary medium, and why certain physical features of Earth were present. Extending outward in the direction of the Sun approximately 40,000 miles, as well as stretching out with a trail away from the Sun to approximately 370,000 miles, the magnetosphere is the area dominated by Earth's strong magnetic field. Measurements by numerous spacecraft have calculated the size and intensity of this field, showing that it contains two belts of very energetic, charged particles. At its upper limits the magnetosphere encounters charged plasma particles thrown off by the Sun, known collectively as the solar wind, which create a boundary where the two come into contact. These electromagnetic fields interact and create shock waves, some of which are responsible for auroral phenomenon such as the aurora borealis or northern lights. The study of the magnetosphere greatly enhanced understanding of the way in which the Sun exerted its influence on this planet.

EXPLORATION OF THE SOLAR SYSTEM. During the decade of the 1960s the U.S. space program began an impressive effort to gather information on the solar system using ground-, air-, and space-based equipment. Especially important in this project was the creation of two types of spacecraft, one a probe that was sent toward a heavenly body, and the second an Earth-orbiting observatory that could gain the clearest resolution available in telescopes because it did not have to contend with the atmosphere. Once again, the compilation of this new data revolutionized humanity's understanding of Earth's immediate environment. Two principal aspects of this research involved activities on a broad front to learn about the Moon, which will be dealt with in a later chapter, and the Sun. One example among the myriad research projects associated with the Sun was the discovery of the solar wind. Although earlier physicists had considered the possibility that the Sun sent not just light and heat into space but also matter, it was not until the advent of the space age that experiments could demonstrate it. University of Chicago professor Eugene N. Parker (1927–) was the leader of a small group of innovative scientists who were attracted to the challenge of observing charged particles and magnetic fields in the solar system. Using such ground-based facilities as the Mount Wilson Observatory; spacecraft such as Lunik 2 (1959),

Explorer 10 (1961), Mariner 2 (1962); and a series of six Orbiting Solar Observatories to provide data, scientific theorists in the decade developed a dynamic model of the solar system that incorporated the solar wind as a fundamental component.

The satellite exploration of the planets, moons, and other bodies of the Solar System also began during the 1960s. Although the most significant findings of this investigation would not come until the 1970s, perhaps the "golden age" of planetary science, studies of the planets captured the imagination of many people from all types of backgrounds like nothing else save the Apollo lunar missions. For all the genuine importance of magnetospheric physics and solar studies, meteorology and plate tectonics, it was photographs of the planets and theories about the origins of the solar system that appealed to a much broader cross-section of the public. As a result NASA had little difficulty in capturing and holding a widespread interest in this aspect of the space science program.

Observation of the planets from Earth-borne instruments had been going on for centuries, and NASA in the 1960s contributed to that through support for radio, radar, and optical astronomy, but the really significant contributions of the space age came from satellites, either probes actually sent to the planets or space-based observatories. These were by far the most costly segments of the planetary investigation program, but they yielded spectacular results from both a scientific and a national prestige standpoint.

A centerpiece of this effort was the Mariner program, originated by NASA in the early part of the decade to investigate the nearby planets. Built by Jet Propulsion Laboratory scientists and technicians, satellites of this program proved enormously productive during the 1960s. The United States claimed the first success in planetary exploration during the summer of 1962 when Mariner 2 was launched toward Venus. In December it arrived at the planet, probing the clouds, estimating planetary temperatures, measuring the charged particle environment, and looking for a magnetic field similar to Earth's magnetosphere (but finding none). In July 1965 Mariner 4 flew by Mars, taking 21 close-up pictures, and Mariner 5 visited Venus in 1967 to investigate the atmosphere. Mariner 6 and Mariner 7, launched in February and March 1969, each passed Mars about five months later, studying its atmosphere and surface to lay the groundwork for an eventual landing on the planet. Among other discoveries from these probes, they found that much of Mars was cratered almost like the Moon, that volcanoes had once been active on the planet, that the frost observed seasonally on the poles was made of carbon dioxide, and that huge plates indicated considerable tectonic activity. Proposals for additional Mariner probes were also considered but because of budgetary considerations did not fly during the decade. These space probes, as well as others not mentioned here, accumulated volumes of data on the near planets and changed many scientific conceptions that had long held sway.

While these successes were great, all was not well with the politics of planetary exploration. In the summer of 1967, even as the technical abilities required to conduct an adventurous space science program were being demonstrated, the planetary science community suffered a devastating defeat in Congress and lost funding for a satellite lander to Mars. No other NASA effort but Project Apollo was more exciting than the Mars program in the middle part of the decade. The planet had long held a special attraction to Americans, so much like Earth and possibly even sustaining life and the lander would have allowed for extended robotic exploration of the Red Planet. A projected $2 billion program, the lander was to use the Saturn V launch vehicle being developed for Apollo. The problem revolved around the lack of consensus among scientists on the validity of the Mars initiative. Some were excited but others thought it was too expensive and placed too many hopes on the shoulders of one project and one project manager. Without that consensus and with other national priorities for spending for "Great Society" social programs, combatting urban unrest, and for the military in Vietnam, the Mars lander was an easy target in Congress. It was the first space science project ever killed on Capitol Hill. The NASA administrator, James E. Webb, frustrated by congressional action and infuriated by internal dissension among scientists, ended all work on planetary probes.

The scientific community learned a hard lesson about the pragmatic, and sometimes brutal, politics associated with the execution of "Big Science" under the suzerainty of the federal government. Most important, it realized that strife within the discipline had to be kept within the discipline in order to put forward a united front against the priorities of other interest groups and other government leaders. While imposing support from the scientific community could not guarantee that any initiative would become a political reality, without it a program could not be funded. It also learned that while a $750 million program found little opposition at any level, a $2 billion project crossed an ill-defined but very real threshold triggering intense competition for those dollars. Having learned these lessons, as well as some more subtle ones, the space science community regrouped and went forward in the latter part of the decade with a trimmed-down Mars lander program, called Viking, that was funded and provided astounding scientific data in the mid-1970s.

INVESTIGATION OF THE UNIVERSE. At the same time that these findings were fundamentally reshaping knowledge of the solar system, space scientists were investigating, and also profoundly affecting, humanity's understanding of the universe beyond. The traditional scientific field of astronomy underwent a tremendous burst of activity in the 1960s because of the ability to study the stars through new types of telescopes. In addition to greatly enhanced capabilities for observation in the visible light spectrum, NASA and other institutions supported the development of a wide range of X-ray, gamma ray, ultraviolet, infrared, microwave, cosmic ray, radar, and radio astronomical projects. These efforts collectively informed the most systematic efforts yet to explain the origins and development of the universe.

Before the space age, all astronomy was performed from the ground and was limited by the Earth's atmosphere that filtered out many types of energy and some visible light sources. The stars and galaxies could only be seen in the visible light and radio spectrums, and large telescopes were constructed to observe at these wavelengths. Using these instruments, in 1927 Edwin P. Hubble (1889–1953), an astronomer at the Mt. Wilson Observatory near Pasadena, California, discovered that other galaxies were apparently receding from the Milky Way, our galaxy, and that the further away they were the faster they retreated. Other astronomers built on Hubble's discovery to create a picture of the universe that had originated at what has been called the "Big Bang" and that has been expanding at a constant rate for 10 to 20 billion years. Alternative theories to the "Big Bang" theory have also been advanced, and much of the exploration of the universe since Hubble has been oriented toward acquiring information that might scientifically prove one or another of these theories. The debate has continued down to the present, and it is one of the truly exciting aspects of space science, bringing humanity into contact with fundamental questions of life and meaning.

The space age provided an opportunity to expand far beyond the capabilities offered by the observatories of Hubble's time. Fundamental to this was the development of a series of Orbiting Astronomical Observatories (OAO), first conceptualized not long after the birth of NASA. Two of these aluminum, octagonally shaped, solar-powered spacecraft were launched during the 1960s. The first failed less than two days into its mission because of a power system failure, but with the launch of OAO 2 on 7 December 1968, the potential of the program began to pay off as it provided an abundance of information on ultraviolet, gamma ray, X-ray, and infrared radiation, on the structure of stars, and on the distribution and density of matter in the interstellar environment. A series of six Orbiting Geophysical Observatories (OGO) also contributed to this study, as well as to the study of the solar system, by taking measurements of cosmic rays, particles and fields in the interplanetary medium as well as radio emissions.

One of the exciting projects in this arena was X-ray astronomy. On 12 June 1962 the first rocket was launched using instruments to detect whether or not X-rays were present in any particular quadrants of the galaxy. It discovered a power source in the center. Calculations demonstrated that X-ray emissions from this source were ten times that of the Sun. In July 1963 another instrument package sent above the atmosphere took readings of the Crab Nebula and found intense X-ray activity emanating from it. In December 1970 the X-ray observatory Uhuru mapped about 85 percent of the sky, then located and measured the intensity of 161 X-ray sources. Many of these turned out to be black holes, a truly significant discovery of a segment of space where mass is so compressed and gravity so great that neither matter nor light can escape. Quantities of X-rays, however, are emitted and can help explain much about the evolution of the universe.

These efforts have been on-going since the beginnings of the space age and represent essential developments in the expansion of human knowledge about the universe. By the early 1970s satellite astronomy had helped to generate a major change in the larger field of astronomy and had reordered thinking on the subject. This occurred in spite of the fact that much of the research was built on the foundations laid by Edwin Hubble and other earlier astronomers. The interplay of energy and matter on a cosmological level has been enormously exciting for many Americans, as findings have been incorporated into the public consciousness. Moreover, although a wide variety of scientific fields enjoyed the yield of the research obtained from the new tools available to scientists, during the decade two important scientific disciplines began to emerge as foremost in the field: the exploration of the solar system and the study of the universe. Throughout this era funding for space science and applications in NASA was never more than $760 million per year (and usually much less), but the return was impressive. The quest for understanding that these efforts helped satisfy gathered momentum during the 1970s as new projects, many of them begun in the 1960s, came to fruition.

ASSESSMENT. In sum, the 1960s saw rapid development both of space science and the technological breakthroughs that made it possible. The result was a radical alteration in the common explanations of the origins and development of the universe. It was a heady environment as funds for space science research rose to about three-quarters of a billion per year in the mid-1960s, about 20 percent of the NASA budget, and satellite probes and orbiters returned far-reaching data that could then be analyzed and incorporated into scientific theories. While much of the history of the American space program has been predicated on the cold war environment of the latter 1950s and the concomitant competition between the superpowers, most but not all of that revolved around human spaceflight initiatives. The centerpiece of this rivalry was Project Apollo, the landing of an American on the Moon before the end of the decade.

CHAPTER 5

TO GO TO THE MOON

KENNEDY AND THE EARLY DEFINITION OF SPACE POLICY. On 25 May 1961 President John F. Kennedy (1917–1963) announced to the nation a goal of sending an American to the Moon before the end of the decade. This decision involved much study and review prior to making it public, and tremendous expenditure and effort to make it a reality by 1969. Only the building of the Panama Canal rivaled the Apollo program's size as the largest nonmilitary technological endeavor ever undertaken by the United States; only the Manhattan Project was comparable in a wartime setting. The human spaceflight imperative was a direct outgrowth of it; Projects Mercury (at least in its latter stages), Gemini, and Apollo were each designed to execute it.

In 1960 John F. Kennedy, a senator from Massachusetts between 1953 and 1960, ran for president as the Democratic candidate, with party wheelhorse Lyndon B. Johnson as his running mate. Using the slogan, "Let's get this country moving again," Kennedy charged the Republican Eisenhower administration with doing nothing about the myriad social, economic, and international problems that festered in the 1950s. He was especially hard on Eisenhower's record in international relations, taking a cold warrior position on a supposed "missile gap" (which turned out not to be the case) wherein the United States lagged far behind the Soviet Union in ICBM technology. He also invoked the cold war rhetoric opposing a communist effort to take over the world and used as his evidence the 1959 revolution in Cuba that brought leftist dictator Fidel Castro to power. The Republican candidate, Richard Nixon (1913–), who had been Eisenhower's vice president (*See Reading No. 8.*), tried to defend his mentor's record but when the results were in Kennedy was elected by a narrow margin of 118,550 out of more than 68 million popular votes cast.

Kennedy as president had little direct interest in the U.S. space program. He was not a visionary enraptured with the romantic image of the last American frontier in space and consumed by the adventure of exploring the unknown. He *was*, on the other hand, a cold warrior with a keen sense of *Realpolitik* in foreign affairs, and worked hard to maintain balance of power and spheres of influence in American/Soviet relations. The Soviet Union's nonmilitary accomplishments in space, therefore, forced Kennedy to respond and to serve notice that the United States was every bit as capable in the space arena as the Soviets. Of course, to prove this fact, Kennedy had to be willing to commit national resources to NASA and the civil space program. The cold war realities of the time, therefore, served as the primary vehicle for an expansion of NASA's activities and for the definition of Project Apollo as the premier civil space effort of the nation. Even more significant, from Kennedy's perspective the cold war necessitated the expansion of the military space program, especially the development of ICBMs and satellite reconnaissance systems.

While Kennedy was preparing to take office, he appointed an ad hoc committee headed by Jerome B. Wiesner (1915–   ) of the Massachusetts Institute of Technology to offer suggestions for American efforts in space. Wiesner, who later headed the President's Science Advisory Committee (PSAC) under Kennedy, concluded that the issue of "national prestige" was too great to allow the Soviet Union leadership in space efforts, and therefore the United States had to enter the field in a substantive way. "Space exploration and exploits," he wrote in a 10 January 1961 report to the president-elect, "have captured the imagination of the peoples of the world. During the next few years the prestige of the United States will in part be determined by the leadership we demonstrate in space activities." Wiesner also emphasized the importance of practical nonmilitary applications of space technology—communications, mapping, and weather satellites among others—and the necessity of keeping up the effort to exploit space for national security through such technologies as ICBMs and reconnaissance satellites. He tended to deemphasize the human spaceflight initiative for very practical reasons. American launch vehicle technology, he argued, was not well developed and the potential of placing an astronaut in space before the Soviets was slim. He thought human spaceflight was a high-risk enterprise with a low-chance of success.

Human spaceflight was also less likely to yield valuable scientific results, and the United States, Wiesner thought, should play to its strength in space science where important results had already been achieved.

Kennedy only accepted part of what Wiesner recommended. He was committed to conducting a more vigorous space program than had been Eisenhower, but he was more interested in human spaceflight than either his predecessor or his science advisor. This was partly because of the drama surrounding Project Mercury and the seven astronauts that NASA was training. Wiesner had cautioned Kennedy about the hyperbole associated with human spaceflight. "Indeed, by having placed the highest national priority on the MERCURY program we have strengthened the popular belief that man in space is the most important aim for our non-military space effort," Wiesner wrote. "The manner in which this program has been publicized in our press has further crystallized such belief." Kennedy, nevertheless, recognized the tremendous public support arising from this program and wanted to ensure that it reflected favorably upon his administration.

But it was a risky enterprise—what if the Soviets were first to send a human into space, what if an astronaut was killed and Mercury was a failure—and the political animal in Kennedy wanted to minimize those risks. The earliest Kennedy pronouncements relative to civil space activity directly addressed these hazards. He offered to cooperate with the Soviet Union, still the only other nation involved in launching satellites, in the exploration of space. In his inaugural address in January 1961 Kennedy spoke directly to Soviet Premier Nikita Khrushchev (1894–1971) and asked him to cooperate in exploring "the stars." In his State of the Union address ten days later, he asked the Soviet Union "to join us in developing a weather prediction program, in a new communications satellite program, and in preparation for probing the distant planets of Mars and Venus, probes which may someday unlock the deepest secrets of the Universe." Kennedy also publicly called for the peaceful use of space, and the limitation of war in that new environment.

In making these overtures Kennedy accomplished several important political ends. First, he appeared to the world as the statesman by seeking friendly cooperation rather than destructive competition with the Soviet Union, knowing full well that there was little likelihood that Khrushchev would accept his offer. Conversely, the Soviets would appear to be monopolizing space for their own personal, and presumably military, benefit. Second, he minimized the goodwill that the Soviet Union enjoyed because of its own success in space vis-à-vis the United States. Finally, if the Soviet Union accepted his call for cooperation, it would tacitly be recognizing the equality of the United States in space activities, something that would also look very good on the world stage.

THE SOVIET CHALLENGE RENEWED. Had the balance of power and prestige between the United States and the Soviet Union remained stable in the spring of 1961, it is quite possible that Kennedy would never have advanced his Moon program and the direction of American space efforts might have taken a radically different course. Kennedy seemed quite happy to allow NASA to execute Project Mercury at a deliberate pace, working toward the orbiting of an astronaut sometime before the middle of the decade, and to build on the satellite programs that were yielding excellent results both in terms of scientific knowledge and practical application. Jerome Wiesner reflected: "If Kennedy could have opted out of a big space program without hurting the country in his judgment, he would have."

Firm evidence for Kennedy's essential unwillingness to commit to a strong space program came in March 1961 when the NASA administrator, James E. Webb (1907–1992), submitted a request that greatly expanded his agency's fiscal year 1962 budget so as to permit a Moon landing before the end of the decade. Kennedy's budget director, David E. Bell (1919–   ), objected to this large increase and told Webb to obtain the president's explicit decision to make the lunar program a part of the administration's effort "to catch up to the Soviet Union in space performance." At White House meetings on 21 and 22 March 1961, Webb and Bell debated the merits of an aggressive lunar landing program before Kennedy and Lyndon Johnson, but in the end the president was unwilling to obligate the nation to a much bigger and more costly space program. Instead, in good political fashion, he approved a modest increase in the NASA budget to allow for development of the big launch vehicles that would eventually be required to support a Moon landing.

A nonchalant pace might have remained the standard for the U.S. civil space effort had not two important events happened that forced Kennedy to act. The Soviet Union's space effort scored another coup on the United States not long after the new president took office. On 12 April 1961 Soviet Cosmonaut Yuri Gagarin (1934–1968) became the first human in space with a one-orbit mission aboard the spacecraft *Vostok 1*. The chance to place a human in space before the Soviets did so had now been lost. The great success of that feat made the gregarious Gagarin a global hero, and he was an effective spokesman for the Soviet Union until his death in an unfortunate aircraft accident. It was only a salve on an open wound, therefore, when Alan Shepard became the first American in space during a 15-minute suborbital flight on 6 May 1961—performing Hugh Dryden's ill-famed "circus stunt" by riding a Redstone booster in his *Freedom 7* Mercury spacecraft.

Comparisons between the Soviet and American flights were inevitable afterwards. Gagarin had flown around the Earth; Shepard had been the cannonball shot from a gun. Gagarin's Vostok spacecraft had weighed 10,428 pounds; *Freedom 7* weighed 2,100 pounds. Gagarin had been weightless for 89 minutes; Shepard for only 5 minutes. "Even though the United States is still the strongest military power and leads in many aspects of the space race," wrote journalist Hanson Baldwin in the *New York Times* not long after Gagarin's flight, "the world—impressed by the spectacular Soviet firsts—believes we lag militarily and technologically." By any unit of measure the United States had not demonstrated technical equality with the Soviet Union, and that fact worried national leaders because of what it would mean in the larger cold war environment. These apparent disparities in technical competence had to be addressed, and Kennedy had to find a way to reestablish the nation's credibility as a technological leader before the world.

Close in the wake of the Gagarin achievement, the Kennedy administration suffered another devastating blow in the cold war that contributed to the sense that action had to be taken. Between 15 and 19 April 1961 the administration supported the abortive Bay of Pigs invasion of Cuba designed to overthrow

Castro. Executed by anti-Castro Cuban refugees armed and trained by the CIA, the invasion was a debacle almost from the beginning. It was predicated on an assumption that the Cuban people would rise up to welcome the invaders and when that proved to be false, the attack could not succeed. American backing of the invasion was a great embarrassment both to Kennedy personally and to his administration. It damaged U.S. relations with foreign nations enormously, and made the communist world look all the more invincible.

While the Bay of Pigs invasion was never mentioned explicitly as a reason for stepping up U.S. efforts in space, the international situation certainly played a role as Kennedy scrambled to recover a measure of national dignity. Wiesner reflected, "I don't think anyone can measure it, but I'm sure it [the invasion] had an impact. I think the President felt some pressure to get something else in the foreground." T. Keith Glennan, NASA administrator under Eisenhower, immediately linked the invasion and the Gagarin flight together as the seminal events leading to Kennedy's announcement of the Apollo decision. He confided in his diary that "In the aftermath of that [Bay of Pigs] fiasco, and because of the successful orbiting of astronauts by the Soviet Union, it is my opinion that Mr. Kennedy asked for a reevaluation of the nation's space program."

REEVALUATING NASA'S PRIORITIES. Two days after the Gagarin flight on 12 April, Kennedy discussed once again the possibility of a lunar landing program with Webb, but the NASA head's conservative estimates of a cost of more than $20 billion for the project was too steep and Kennedy delayed making a decision. A week later, at the time of the Bay of Pigs invasion, Kennedy called Johnson, who headed the National Aeronautics and Space Council, to the White House to discuss strategy for catching up with the Soviets in space. Johnson agreed to take the matter up with the Space Council and to recommend a course of action. It is likely that one of the explicit programs that Kennedy asked Johnson to consider was a lunar landing program, for the next day, 20 April 1961, he followed up with a memorandum to Johnson raising fundamental questions about the project. Kennedy wanted to know if "we have a chance of beating the Soviets by . . . a trip around the moon, or by a rocket to land on the moon, or by a rocket to go to the moon and back with a man. Is there any other space program which promises dramatic results in which we could win?" (*See Reading No. 9.*)

While he waited for the results of Johnson's investigation, this memo made it clear that Kennedy had a pretty good idea of what he wanted to do in space. He confided in a press conference on 21 April that he was leaning toward committing the nation to a large-scale project to land Americans on the Moon. "If we can get to the moon before the Russians, then we should," he said, adding that he had asked his vice president to review options for the space program. This was the first and last time that Kennedy said anything in public about a lunar landing program until he officially unveiled the plan. It is also clear that Kennedy approached the lunar landing effort essentially as a response to the competition between the United States and the Soviet Union. For Kennedy the Moon landing program, conducted in the tense cold war environment of the early 1960s, was a strategic decision directed toward advancing the far-flung interests of the United States in the international arena. It aimed toward recapturing the prestige that the nation had lost as a result of Soviet successes and U.S. failures. It was, as political scientist John M. Logsdon has suggested, "one of the last major political acts of the Cold War. The Moon Project was chosen to symbolize U.S. strength in the head-to-head global competition with the Soviet Union."

Lyndon Johnson probably understood these circumstances very well, and for the next two weeks his Space Council diligently considered, among other possibilities, a lunar landing before the Soviets. As early as 22 April, NASA's Hugh Dryden had responded to a request for information from the Space Council about a Moon program by writing that there was "a chance for the U.S. to be the first to land a man on the moon and return him to earth if a determined national effort is made." He added that the earliest this feat could be accomplished was 1967, but that to do so would cost about $33 billion dollars, a figure $10 billion more than the whole projected NASA budget for the next ten years. A week later Wernher von Braun, director of NASA's George C. Marshall Space Flight Center at Huntsville,

Alabama, and head of the big booster program needed for the lunar effort, responded to a similar request for information from Johnson. He told the vice president that "we have a sporting chance of sending a 3-man crew *around the moon* ahead of the Soviets" and "an excellent chance of beating the Soviets to the *first landing of a crew on the moon* (including return capability, of course.)" He added that "with an all-out crash program" the United States could achieve a landing by 1967 or 1968.

After gaining these technical opinions, Johnson began to poll political leaders for their sense of the propriety of committing the nation to an accelerated space program with Project Apollo as its centerpiece. He brought in Senators Robert Kerr (D-OK) (1896–1963) and Styles Bridges (R-NH) (1898–1961) and spoke with several representatives to ascertain if they were willing to support an accelerated space program. While only a few were hesitant, Robert Kerr worked to allay their concerns. He called on James Webb, who had worked for his business conglomerate during the 1950s, to give him a straight answer about the project's feasibility. Kerr told his congressional colleagues that Webb was enthusiastic about the program and "that if Jim Webb says we can a land a man on the moon and bring him safely home, then it can be done." This endorsement secured considerable political support for the lunar project. Johnson also met with several businessmen and representatives from the aerospace industry and other government agencies to ascertain the consensus of support for a new space initiative. Most of them also expressed support.

Air Force General Bernard A. Schriever, commander of the Air Force Systems Command that developed new technologies, expressed the sentiment of many people by suggesting that an accelerated lunar landing effort "would put a focus on our space program." He believed it was important for the United States to build international prestige and that the return was more than worth the price to be paid. Secretary of State Dean Rusk (1909–   ), a member of the Space Council, was also a supporter of the initiative because of the Soviet Union's image in the world. He wrote to the Senate Space Committee a little later that "We must respond to their conditions; otherwise we risk a basic misunderstanding on the part of the uncommitted countries, the Soviet Union, and possibly our allies concerning the direction in which power is moving and where long-term advantage lies." It was clear early in these deliberations that Johnson was in favor of an expanded space program in general and a maximum effort to land an astronaut on the Moon. Whenever he heard reservations Johnson used his forceful personality to persuade. "Now," he asked, "would you rather have us be a second-rate nation or should we spend a little money?"

In an interim report to the president on 28 April 1961, Johnson concluded that "The U.S. can, if it will, firm up its objectives and employ its resources with a reasonable chance of attaining world leadership in space during this decade," and recommended committing the nation to a lunar landing. (*See Reading No. 10.*) In this exercise Johnson had built, as Kennedy had wanted, a strong justification for undertaking Project Apollo but he had also moved on to develop a greater consensus for the objective among key government and business leaders.

THE NASA POSITION. While NASA's leaders were generally pleased with the course Johnson was recommending—they recognized and mostly agreed with the political reasons for adopting an aggressive lunar landing program—they wanted to shape it as much as possible to the agency's particular priorities. NASA Administrator James Webb, well known as a skilled political leader who could seize an opportunity, organized a short-term effort to accelerate and expand a long-range NASA master plan for space exploration. A fundamental part of this effort addressed a legitimate concern that the scientific and technological advancements for which NASA had been created not be eclipsed by the political necessities of international rivalries. Webb conveyed the concern of the agency's technical and scientific community to Jerome Wiesner on 2 May 1961, noting that "the most careful consideration must be given to the scientific and technological components of the total program and how to present the picture to the world and to our own nation of a program that has real value and validity and from which solid additions to knowledge can be made, even if every one of the specific so-called 'spectacular' flights or events are done after they have been accomplished by the Russians." He asked that Wiesner help him "make sure that this component of solid, and yet imaginative, total scientific and technological value is built in."

Partly in response to this concern, Johnson asked NASA to provide for him a set of specific recommendations on how a scientifically viable Project Apollo, would be accomplished by the end of the decade. What emerged was a comprehensive space policy planning document that had the lunar landing as its centerpiece but that attached several ancillary funding items to enhance the program's scientific value and advance space exploration on a broad front:

1. spacecraft and boosters for the human flight to the Moon
2. scientific satellite probes to survey the Moon
3. a nuclear rocket
4. satellites for global communications
5. satellites for weather observation
6. scientific projects for Apollo landings

Johnson accepted these recommendations and passed them to Kennedy who approved the overall plan.

The last major area of concern was the timing for the Moon landing. The original NASA estimates had given a target date of 1967, but as the project became more crystallized agency leaders recommended not committing to such a strict deadline. James Webb, realizing the problems associated with meeting target dates based on NASA's experience in spaceflight, suggested that the president commit to a landing by the end of the decade, giving the agency another two years to solve any problems that might arise. The White House accepted this proposal.

DECISION. President Kennedy unveiled the commitment to execute Project Apollo on 25 May 1961 in a speech on "Urgent National Needs," billed as a second State of the Union message. He told Congress that the United States faced extraordinary challenges and needed to respond extraordinarily. In announcing the lunar landing commitment he said:

If we are to win the battle that is going on around the world between freedom and tyranny, if we are to win the battle for men's minds, the dramatic achievements in space which occurred in recent weeks should have made clear to us all, as did the Sputnik in 1957, the impact of this adventure on the minds of men everywhere who are attempting to make a determination of which road they should take. . . . We go into space because whatever mankind must undertake, free men must fully share.

Then he added: "I believe this Nation should commit itself to achieving the goal, before this decade is out, of landing a man on the moon and returning him safely to earth. No single space project in this period will be more impressive to mankind, or more important for the long-range exploration of space; and none will be so difficult or expensive to accomplish." (*See Reading No. 11.*)

ASSESSMENT. The president had correctly gauged the mood of the nation. His commitment captured the American imagination and was met with overwhelming support. No one seemed concerned either about the difficulty or about the expense at the time. Congressional debate was perfunctory and NASA found itself literally pressing to expend the funds committed to it during the early 1960s. Like most political decisions, at least in the U.S. experience, the decision to carry out Project Apollo was an effort to deal with an unsatisfactory situation (world perception of Soviet leadership in space and technology). As such Apollo was a remedial action ministering to a variety of political and emotional needs floating in the ether of world opinion. Apollo addressed these problems very well, and was a worthwhile action if measured only in those terms. In announcing Project Apollo, Kennedy put the world on notice that the United States would not take a back seat to its superpower rival. John Logsdon commented: "By entering the race with such a visible and dramatic commitment, the United States effectively undercut Soviet space spectaculars without doing much except announcing its intention to join the contest." It was an effective symbol, just as Kennedy had intended.

It also gave the United States an opportunity to shine. The lunar landing was so far beyond the capabilities of either the United States or the Soviet Union in 1961 that the early lead in space activities taken by the Soviets would not predetermine the outcome. It gave the United States a reasonable chance of overtaking the Soviet Union in space activities and recovering a measure of lost status.

Even though Kennedy's political objectives were essentially achieved with the decision to go to the Moon, there were other aspects of the Apollo commitment that require assessment. Those who wanted to see a vigorous space program, a group led by NASA scientists and engineers, obtained their wish with Kennedy's announcement. An opening was available to this group in 1961 that had not existed at any time during the Eisenhower administration, and they made the most of it. They inserted into the overall package supporting Apollo additional programs that they believed would greatly strengthen the scientific and technological return on the investment to go to the Moon. In addition to seeking international prestige, this group proposed an accelerated and integrated national space effort incorporating both scientific and commercial components.

A unique confluence of political necessity, personal commitment and activism, scientific and technological ability, economic prosperity, and public mood made possible the 1961 decision to carry out an aggressive lunar landing program. It then fell to NASA and other organizations of the federal government to accomplish the task set out in a few short paragraphs by the president. By the time that the goal would be accomplished in 1969, few of the key figures associated with the decision would still be in leadership positions in the government. Kennedy fell victim to an assassin's bullet in 1963, and Wiesner returned to MIT soon afterwards. Lyndon Johnson, of course, succeeded Kennedy as president but left office in January 1969 just a few months before the first lunar landing. Webb resolutely guided NASA through most of the 1960s, but his image was tarnished by, among other things, a 1967 Apollo accident that killed three astronauts, and he retired from office under something of a cloud in October 1968. Hugh Dryden and several early supporters of Apollo in Congress died during the 1960s and never saw the program successfully completed.

CHAPTER 6

MANAGING THE MOON PROGRAM

GEARING UP FOR PROJECT APOLLO. When John F. Kennedy told his audience in 1961 that he was committing the nation to a lunar landing he probably had little understanding of the technological, scientific, and financial energy that his decision unleashed. When NASA learned of the president's decision to move forward with the Moon landing, its leaders reacted with the mixed emotions of excitement and anguish. They could finally realize many of their dreams for an aggressive space program. At the same time, it placed an enormous burden on the agency. How would NASA go about accomplishing the president's goal? The technological challenge was enormous, and directing the Apollo effort required a special genius for organization and management.

HARNESSING RESOURCES. The first challenge NASA leaders faced was securing funding. There was an old NACA truism that NASA co-opted during the Apollo era: "no bucks, no Buck Rogers." While the truism was only partially correct, definition of costs for the project was critical to the development of a viable lunar landing program. While Congress enthusiastically appropriated funding for Apollo after the president's mandate, NASA Administrator James E. Webb was rightly concerned that the momentary sense of crisis would subside and that the political consensus present for Apollo in 1961 would abate. He tried, albeit without much success, to lock the presidency and the Congress into a long-term obligation to support the program. While they had made an intellectual commitment, NASA's leadership was concerned that they might renege on the economic part of the bargain at some future date. (*See Reading No. 12.*)

Initial NASA estimates of the costs of Project Apollo were about $20 billion through the end of the decade, a figure approaching $150 billion in 1994 dollars when accounting for inflation. Webb quickly stretched those initial estimates for

Apollo as far as possible, with the intent that even if NASA did not receive its full budget requests, as it did not during the latter half of the decade, it would still be able to complete Apollo. At one point in 1963, for instance, Webb came forward with a NASA funding projection through 1970 for more than $35 billion. As it turned out Webb was able to sustain the momentum of Apollo through the decade, largely because of his rapport with key members of Congress and with Lyndon B. Johnson, who became president in November 1963.

Project Apollo, backed by sufficient funding, was the tangible result of an early national commitment in response to a perceived threat to the United States by the Soviet Union. NASA leaders recognized that while the size of the task was enormous, it was still technologically and financially within their grasp, but they had to move forward quickly. Accordingly, the space agency's annual budget increased from $500 million in 1960 to a high point of $5.2 billion in 1965. The NASA funding level represented 5.3 percent of the federal budget in 1965. A comparable percentage of the $1.23 trillion federal budget in 1992 would have equaled more than $65 billion for NASA, whereas the agency's actual budget then stood at less than $15 billion.

Out of the budgets appropriated for NASA each year approximately 50 percent went directly for human spaceflight, and the vast majority of that went directly toward Apollo. Between 1959 and 1973 NASA spent $23.6 billion on human spaceflight, exclusive of infrastructure and support, of which nearly $20 billion was for Apollo. In addition, Webb sought to expand the definition of Project Apollo beyond just the mission of landing humans on the Moon. As a result even those projects not officially funded under the Apollo line item could be justified as supporting the mission.

Webb took this approach for essentially two reasons. First, he and the rest of the NASA leadership were committed to developing a broad-based space exploration program, not the execution of a single project activity, even one as far-reaching as Apollo. Unfortunately, because of the massive size of the lunar landing mission, and especially because of its cost, there was little opportunity to undertake additional large space exploration initiatives. Using Apollo as a vehicle to accomplish a host of scientific and technical activities was a practical solution to the funding problem. Second, by attaching broad scientific and technical enterprises to Project Apollo, NASA incorporated various other groups and activities into the program and helped to elicit the continued support of those involved. While not entirely successful in this effort, the result was that much additional space science, education, and a host of other activities were carried out under the rubric of Apollo than might have been accomplished otherwise.

As a result of these strategies, Webb tried to hang every possible NASA program on the lunar mission so as to ensure its funding. For example, he argued that many scientific missions were required to support Apollo and the program, therefore, became an umbrella for such Moon probes as the Ranger, Lunar Orbiter, and Surveyor series. Since Apollo required a system of radar tracking, telemetry, and communications NASA justified its development on the basis of the lunar mission, although it was used for much more than the landing project. Webb also launched a broad-based educational effort to train aerospace engineers and scientists. Using the hook of the lunar landing's need for great advances in science and technology, he sought to expand American education and research by channeling millions of dollars into the nation's educational institutions via Apollo. The centerpiece of this effort was the Sustaining University Program inaugurated in 1962 in the name of Apollo. Accordingly, by 1970 NASA had paid the bills for the graduate educations of more than 5,000 scientists and engineers at a cost of over $100 million. It had also spent more than $32 million on the construction of university laboratories and given more than $50 million worth of multidisciplinary grants to some 50 universities.

For seven years after Kennedy's Apollo decision, through October 1968, James Webb politicked, coaxed, cajoled, and maneuvered for NASA in Washington. A long-time Washington insider—the former director of the Bureau of the Budget and under secretary of state during the Truman administration—he was a master at bureaucratic politics, understanding that it was essentially a system of mutual give and take. For instance, while the native North Carolinian may also have genuinely believed in the Johnson administration's Civil Rights bill that went before Congress in 1964, as a personal favor to the president he lobbied for its passage on Capitol Hill. This secured for him Johnson's gratitude, which he then used to secure the administration's backing of NASA's initiatives. In addition, Webb wielded the money appropriated for Apollo to build up a constituency for NASA that was both powerful and vocal. This type of gritty pragmatism also characterized Webb's dealings with other government officials and members of Congress throughout his tenure as administrator. When give and take did not work, as was the case on occasion with some members of Congress, Webb used the presidential directive as a hammer to get his way. Usually this proved successful. After Kennedy's assassination in 1963, moreover, he sometimes appealed for continued political support for Apollo because it represented a fitting tribute to the fallen leader. In the end, through a variety of methods Administrator Webb built a seamless web of political liaisons that brought continued support for and resources to accomplish the Apollo Moon landing on the schedule Kennedy had announced.

Funding was not the only critical component for Project Apollo. To realize the goal of Apollo under the strict time constraints mandated by the president, personnel had to be mobilized. A recent study estimates that approximately 1 in 50 Americans worked on some aspect of the Apollo program during its existence. They might not have been so closely involved that they recognized a direct linkage, and certainly not all supported it at any given time, but Apollo's presence was pervasive in American business and society. This took two forms. First, by 1966 the agency's civil service rolls had grown to 36,000 people from the 10,000 employed at NASA in 1960. Additionally, NASA's leaders made an early decision that they would have to rely upon outside researchers and technicians to complete Apollo, and contractor employees working on the program increased by a factor of 10, from 36,500 in 1960 to 376,700 in 1965. Private industry, research institutions, and universities, therefore, provided the majority of personnel working on Apollo.

To incorporate the great amount of work undertaken for the project into the formal bureaucracy never seemed a particularly savvy idea, and as a result during the 1960s somewhere between 80 and 90 percent of NASA's overall budget went for contracts to purchase goods and services from others. Although the magnitude of the endeavor had been much smaller than with Apollo, this reliance on the private sector and universities for the bulk of the effort originated early in NASA's history under T. Keith Glennan, in part because of the Eisenhower administration's mistrust of large government establishments. Although neither Glennan's successor, nor Kennedy shared that mistrust, they found that it was both good politics and the best way of getting Apollo done on the presidentially approved schedule. It was also very nearly the only way to harness talent and institutional resources already in existence in the emerging aerospace industry and the country's leading research universities.

In addition to these other resources, NASA moved quickly during the early 1960s to expand its physical capacity so that it could accomplish Apollo. In 1960 the space agency consisted of a small headquarters in Washington, its three inherited NACA research centers, the Jet Propulsion Laboratory, the Goddard Space Flight Center, and the Marshall Space Flight Center. With the advent of Apollo, these installations grew rapidly. In addition, NASA added three new facilities specifically to meet the demands of the lunar landing program. In 1962 it created the Manned Spacecraft Center (renamed the Lyndon B. Johnson Space Center in 1973), near Houston, Texas, to design the Apollo spacecraft and the launch platform for the lunar lander. This center also became the home of NASA's astronauts and the site of mission control. NASA then greatly expanded for Apollo the Launch Operations Center at Cape Canaveral on Florida's eastern seacoast. Renamed the John F. Kennedy Space Center on 29 November 1963, this installation's massive and expensive Launch Complex 39 was the site of all Apollo lunar firings. Additionally, the spaceport's Vehicle Assembly Building was a huge and expensive 36-story structure where the Saturn/Apollo rockets were assembled. Finally, to support the development of the Saturn launch vehicle, in October 1961 NASA created on a deep south bayou the Mississippi Test Facility, renamed the John C. Stennis Space Center in 1988. The cost of this expansion was great, more than 2.2 billion over the decade, with 90 percent of it expended before 1966.

THE PROGRAM MANAGEMENT CONCEPT. The mobilization of resources was not the only challenge facing those charged with meeting President Kennedy's goal. NASA had to meld disparate institutional cultures and approaches into an inclusive organization moving along a single unified path. Each NASA installation, university, contractor, and research facility had differing perspectives on how to go about the task of accomplishing Apollo. To bring a semblance of order to the program, NASA expanded the "program management" concept borrowed by T. Keith Glennan in the late 1950s from the military/industrial complex, bringing in military managers to oversee Apollo. The central figure in this process was U.S. Air Force Major General Samuel C. Phillips (1921–1990), the architect of the Minuteman ICBM program before coming to NASA in 1962. Answering directly to the Office of Manned Space Flight at NASA headquarters, which in turn reported to the NASA administrator, Phillips created an omnipotent program office with centralized authority over design, engineering, procurement, testing, construction, manufacturing, spare parts, logistics, training, and operations.

One of the fundamental tenets of the program management concept was that three critical factors—cost, schedule, and reliability—were interrelated and had to be managed as a group. Many also recognized these factor's constancy; if program managers held cost to a specific level, then one of the other two factors, or both of them to a somewhat lesser degree, would be adversely affected. This held true for the Apollo program. The schedule, dictated by the president, was firm. Since humans were involved in the flights, and since the president had directed that the lunar landing be conducted safely, the program managers placed a heavy emphasis on reliability. Accordingly, Apollo used redundant systems extensively so that failures would be both predictable and minor in result. The significance of both of these factors forced the third factor, cost, much higher than might have been the case with a more leisurely lunar program such as had been conceptualized in the latter 1950s. As it was, this was the price paid for success under the Kennedy mandate and program managers made conscious decisions based on a knowledge of these factors.

The program management concept was recognized as a critical component of Project Apollo's success in November 1968, when *Science* magazine, the publication of the American Association for the Advancement of Science, observed:

In terms of numbers of dollars or of men, NASA has not been our largest national undertaking, but in terms of complexity, rate of growth, and technological sophistication it has been unique. . . . It may turn out that [the space program's] most valuable spin-off of all will be human rather than technological: better knowledge of how to plan, coordinate, and monitor the multitudinous and varied activities of the organizations required to accomplish great social undertakings.

Understanding the management of complex structures for the successful completion of a multifarious task was an important outgrowth of the Apollo effort.

This management concept under Phillips involved more than 500 contractors working on both large and small aspects of Apollo. For example, the prime contracts awarded to industry for the principal components of just the Saturn V included the Boeing Company for the S-IC, first stage; North American Aviation, S-II, second stage; the Douglas Aircraft Corporation, S-IVB, third stage; the Rocketdyne Division of North American Aviation, J-2 and F-1 engines; and International Business Machines (IBM), Saturn instruments. These prime contractors, with more than 250 subcontractors, provided millions of parts and components for use in the Saturn launch vehicle, all meeting exacting specifications for performance and reliability. The total cost expended on development of the Saturn launch vehicle was massive, amounting to $9.3 billion. So huge was the overall Apollo endeavor that NASA's procurement actions rose from roughly 44,000 in 1960 to almost 300,000 by 1965.

Getting all of the personnel elements to work together challenged the program managers, regardless of whether they were civil service, industry, or university personnel. There were various communities within NASA that differed over priorities and competed for resources. The two most identifiable groups were the engineers and the scientists. As ideal types, engineers usually worked in teams to build hardware that could carry out the missions necessary to a successful Moon landing by the end of the decade. Their primary goal involved building vehicles that would function reliably within the fiscal resources allocated to Apollo. Again as ideal types, space scientists engaged in pure research and were more concerned with designing experiments that would expand scientific knowledge about the Moon. They also tended to be individualists, unaccustomed to regimentation and unwilling to concede gladly the direction of projects to outside entities. The two groups contended with each other over a great variety of issues associated with Apollo. For instance, the scientists disliked having to configure payloads so that they could meet time, money, or launch vehicle constraints. The engineers, likewise, resented changes to scientific packages added after project definition because these threw their hardware efforts out of kilter. Both had valid complaints and had to maintain an uneasy cooperation to accomplish Project Apollo.

The scientific and engineering communities within NASA, also, were not monolithic, and differences among them thrived. Add to these groups representatives from industry, universities, and research facilities, and competition on all levels to further their own scientific and technical areas was the result. The NASA leadership generally viewed this pluralism as a positive force within the space program, for it ensured that all sides aired their views and emphasized the honing of positions to a fine edge. Competition, most people concluded, made for a more precise and viable space exploration effort. There were winners and losers in this strife, however, and sometimes ill-will was harbored for years. Moreover, if the conflict became too great and spilled into areas where it was misunderstood, it could be devastating to the conduct of the lunar program. The head of the Apollo program worked hard to keep these factors balanced and to promote order so that NASA could accomplish the presidential directive.

Another important management issue arose from the agency's inherited culture of in-house research. Because of the magnitude of Project Apollo, and its time schedule, most of the nitty-gritty work had to be done outside NASA by means of contracts. As a result, with a few important exceptions, NASA scientists and engineers did not build flight hardware, or even operate missions. Rather, they planned the program, prepared guidelines for execution, competed contracts, and oversaw work accomplished elsewhere. This grated on those NASA personnel oriented toward research, and prompted disagreements over how to carry out the lunar landing goal. Of course, they had reason for complaint beyond the simplistic argument of wanting to be "dirty-handed" engineers; they had to have enough in-house expertise to ensure program accomplishment. If scientists or engineers did not have a professional competence on a par with the individuals actually doing the work, how could they oversee contractors creating the hardware and performing the experiments necessary to meet the rigors of the mission?

One anecdote illustrates this point. The Saturn second stage was built by North American Aviation at its plant at Seal Beach, California, shipped to NASA's Marshall Space Flight Center, Huntsville, Alabama, and there tested to ensure that it met contract specifications. Problems developed on this piece of the Saturn effort and Wernher von Braun began intensive investigations. Essentially his engineers completely disassembled and examined every part of every stage delivered by North American to ensure no defects. This was an enormously expensive and time-consuming process, grinding the stage's production schedule almost to a standstill and jeopardizing the presidential timetable.

When this happened Webb told von Braun to desist, adding that "We've got to trust American industry." The issue came to a showdown at a meeting where the Marshall rocket team was asked to explain its extreme measures. While doing so, one of the engineers produced a rag and told Webb that "this is what we find in this stuff." The contractors, the Marshall engineers believed, required extensive oversight to ensure they produced the highest quality work. A compromise emerged that was called the 10 percent rule: 10 percent of all funding for NASA was to be spent to ensure in-house expertise and in the process check contractor reliability.

HOW DO WE GO TO THE MOON? One of the critical early management decisions made by NASA was the method of going to the Moon. No controversy in Project Apollo more significantly caught up the tenor of competing constituencies in NASA than this one. There were three basic approaches that were advanced to accomplish the lunar mission:

1. *Direct Ascent* called for the construction of a huge booster that launched a spacecraft, sent it on a course directly to the Moon, landed a large vehicle, and sent some part of it back to Earth. The Nova booster project, which was to have been capable of generating up to 40 million pounds of thrust, would have been able to accomplish this feat. Even if other

factors had not impaired the possibility of direct ascent, the huge cost and technological sophistication of the Nova rocket quickly ruled out the option and resulted in cancellation of the project early in the 1960s despite the conceptual simplicity of the direct ascent method. The method had few advocates when serious planning for Apollo began.

2.   *Earth-Orbit Rendezvous* was the logical first alternative to the direct ascent approach. It called for the launching of various modules required for the Moon trip into an orbit above the Earth, where they would rendezvous, be assembled into a single system, refueled, and sent to the Moon. This could be accomplished using the Saturn launch vehicle already under development by NASA and capable of generating 7.5 million pounds of thrust. A logical component of this approach was also the establishment of a space station in Earth orbit to serve as the lunar mission's rendezvous, assembly, and refueling point. In part because of this prospect, a space station emerged as part of the long-term planning of NASA as a jumping-off place for the exploration of space. This method of reaching the Moon, however, was also fraught with challenges, notably finding methods of maneuvering and rendezvousing in space, assembling components in a weightless environment, and safely refueling spacecraft.

3.   *Lunar-Orbit Rendezvous* proposed sending the entire lunar spacecraft up in one launch. It would head to the Moon, enter into orbit, and dispatch a small lander to the lunar surface. It was the simplest of the three methods, both in terms of development and operational costs, but it was risky. Since rendezvous was taking place in lunar, instead of Earth, orbit there was no room for error or the crew could not get home. Moreover, some of the trickiest course corrections and maneuvers had to be done after the spacecraft had been committed to a circumlunar flight. The Earth-orbit rendezvous approach kept all the options for the mission open longer than the lunar-orbit rendezvous mode.

Inside NASA, advocates of the various approaches contended over the method of flying to the Moon while the all-important clock that Kennedy had started continued to tick. It was critical that a decision not be delayed, because the mode of flight in part dictated the spacecraft developed. While NASA engineers could proceed with building a launch vehicle, the Saturn, and define the basic components of the spacecraft—a habitable crew compartment, a baggage car of some type, and a jettisonable service module containing propulsion and other expendable systems— they could not proceed much beyond rudimentary conceptions without a mode decision. The NASA Rendezvous Panel at Langley Research Center, headed by John C. Houbolt (1919–   ), pressed hard for the lunar-orbit rendezvous as the most expeditious means of accomplishing the mission. Using sophisticated technical and economic arguments, over a period of months in 1961 and 1962 Houbolt's group advocated and persuaded the rest of NASA's leadership that lunar-orbit rendezvous was not the risky proposition that it had earlier seemed.

The last to give in was Wernher von Braun and his associates at the Marshall Space Flight Center. This group favored the Earth-orbit rendezvous because the direct ascent approach was technologically unfeasible before the end of the 1960s, because it provided a logical rationale for a space station, and because it ensured an extension of the Marshall workload (something that was always important to center directors competing inside the agency for personnel and other resources). At an all-day meeting on 7 June 1962 at Marshall, NASA leaders met to hash out these differences, with the debate getting heated at times. After more than six hours of discussion von Braun finally gave in to the lunar-orbit rendezvous mode, saying that its advocates had demonstrated adequately its feasibility and that any further contention would jeopardize the president's timetable.

With internal dissension quieted, NASA moved to announce the Moon landing mode to the public in the summer of 1962. As it prepared to do so, however, Kennedy's science adviser, Jerome B. Wiesner, raised objections because of the inherent risk it brought to the crew. As a result of this opposition, Webb backpedaled and stated that the decision was tentative and that NASA would sponsor further studies. The issue reached a climax at the Marshall Space Flight Center in September 1962 when President Kennedy, Wiesner, Webb, and several other Washington figures visited von Braun. As the entourage viewed a mock-up of a Saturn V first stage booster during a photo opportunity for the media, Kennedy nonchalantly mentioned to von Braun, "I understand you and Jerry disagree about the right

way to go to the moon." Von Braun acknowledged this disagree-
ment, but when Wiesner began to explain his concern, Webb,
who had been quiet until this point, began to argue with him
"for being on the wrong side of the issue." While the mode
decision had been an uninteresting technical issue before, it
then became a political concern hashed over in the press for days
thereafter. The science advisor to British Prime Minister Harold
Macmillan, who had accompanied Wiesner on the trip, later
asked Kennedy on Air Force One how the debate would turn
out. The president told him that Wiesner would lose, "Webb's
got all the money, and Jerry's only got me." Kennedy was right,
Webb lined up political support in Washington for the lunar-orbit
rendezvous mode and announced it as a final decision on 7
November 1962. This set the stage for the operational aspects
of Apollo.

CHAPTER 7

A SPRINT TO THE MOON

CONDUCTING THE MERCURY PROGRAM. At the time of the announcement of Project Apollo by President Kennedy in May 1961 NASA was still consumed with the task of placing an American in orbit through Project Mercury. Stubborn problems arose, however, seemingly at every turn. The first space flight of an astronaut, made by Alan B. Shepard, had been postponed for weeks so NASA engineers could resolve numerous details and only took place on 6 May 1961, less than three weeks before the Apollo announcement. The second flight, a suborbital mission like Shepard's, launched on 21 July 1961, also had problems. The hatch blew off prematurely from the Mercury capsule, *Liberty Bell 7*, and it sank into the Atlantic Ocean before it could be recovered. In the process the astronaut, "Gus" Grissom, nearly drowned before being hoisted to safety in a helicopter. These suborbital flights, however, proved valuable for NASA technicians who found ways to solve or work around literally thousands of obstacles to successful space flight.

As these issues were being resolved, NASA engineers began final preparations for the orbital aspects of Project Mercury. In this phase NASA planned to use a Mercury capsule capable of supporting a human in space for not just minutes, but eventually for as much as three days. As a launch vehicle for this Mercury capsule, NASA used the more powerful Atlas instead of the Redstone. But this decision was not without controversy. There were technical difficulties to be overcome in mating it to the Mercury capsule to be sure, but the biggest complication was a debate among NASA engineers over its propriety for human spaceflight.

When first conceived in the 1950s many believed Atlas was a high-risk proposition because to reduce its weight Convair Corp. engineers under the direction of Karel J. Bossart (1904–1975), a pre-World War II immigrant from Belgium, designed the booster with a very thin, internally pressurized fuselage instead of massive struts and a thick metal skin. The "steel balloon," as it was sometimes called, employed techniques that ran counter to a conservative engineering approach used by Wernher von Braun for the V-2 and the Redstone at Huntsville, Alabama. Von Braun, according to Bossart, needlessly designed his boosters like "bridges," to withstand any possible shock. For his part, von Braun thought the Atlas too flimsy to hold up during launch. He considered Bossart's approach much too dangerous for human spaceflight, remarking that the astronaut using the "contraption," as he called the Atlas booster, "should be getting a medal just for sitting on top of it before he takes off!" The reservations began to melt away, however, when Bossart's team pressurized one of the boosters and dared one of von Braun's engineers to knock a hole in it with a sledge hammer. The blow left the booster unharmed, but the recoil from the hammer nearly clubbed the engineer.

Most of the differences had been resolved by the first successful orbital flight of an unoccupied Mercury-Atlas combination in September 1961. On 29 November the final test flight took place, this time with the chimpanzee Enos occupying the capsule for a two-orbit ride before being successfully recovered in an ocean landing. Not until 20 February 1962, however, could NASA get ready for an orbital flight with an astronaut. On that date John Glenn became the first American to circle the Earth, making three orbits in his *Friendship 7* Mercury spacecraft. The flight was not without problems, however; Glenn flew parts of the last two orbits manually because of an autopilot failure and left his normally jettisoned retrorocket pack attached to his capsule during reentry because of a loose heat shield.

Glenn's flight provided a healthy increase in national pride, making up for at least some of the earlier Soviet successes. The public, more than celebrating the technological success, embraced Glenn as a personification of heroism and dignity. Hundreds of requests for personal appearances by Glenn poured into NASA headquarters, and NASA learned much about the power of the astronauts to sway public opinion. The NASA leadership made Glenn available to speak at some events, but more often substituted other astronauts and declined many other invitations. Among other engagements, Glenn did address a joint session of Congress and participated in several ticker-tape parades around the country. NASA discovered in the process of this hoopla a powerful public relations tool that it has employed ever since.

Three more successful Mercury flights took place during 1962 and 1963. Scott Carpenter made three orbits on 20 May 1962, and on 3 October 1962 Walter Schirra flew six orbits. The capstone of Project Mercury was the 15–16 May 1963 flight of Gordon Cooper, who circled the Earth 22 times in 34 hours. The program had succeeded in accomplishing its purpose: to successfully orbit a human in space, explore aspects of tracking and control, and to learn about microgravity and other biomedical issues associated with spaceflight.

PROJECT GEMINI. Even as the Mercury program was underway and work took place developing Apollo hardware, NASA program managers perceived a huge gap in the capability for human spaceflight between that acquired with Mercury and what would be required for a lunar landing. They closed most of the gap by experimenting and training on the ground, but some issues required experience in space. Three major areas immediately arose where this was the case. The first was the ability in space to locate, maneuver toward, and rendezvous and dock with another spacecraft. The second was closely related, the ability of astronauts to work outside a spacecraft. The third involved the collection of more sophisticated physiological data about the human response to extended spaceflight.

To gain experience in these areas before Apollo could be readied for flight, NASA devised Project Gemini. Hatched in the fall of 1961 by engineers at Robert Gilruth's Space Task Group in cooperation with McDonnell Aircraft Corp. technicians, builders of the Mercury spacecraft, Gemini started as a larger Mercury Mark II capsule but soon became a totally different proposition. It could accommodate two astronauts for extended flights of more than two weeks. It pioneered the use of fuel cells instead of batteries to power the ship, and incorporated a series of modifications to hardware. Its designers also toyed with the possibility of using a paraglider being developed at Langley Research Center for "dry" landings instead of a "splashdown" in water and recovery by the Navy. The whole system was to be powered by the newly developed Titan II launch vehicle, another ballistic missile developed for the Air Force. A central reason for this program was to perfect techniques for rendezvous and docking, so NASA appropriated from the military some Agena rocket upper stages and fitted them with docking adapters.

Problems with the Gemini program abounded from the start. The Titan II had longitudinal oscillations, called the "pogo" effect because it resembled the behavior of a human on a pogo stick (then a popular toy). Overcoming this problem required engineering imagination and long hours of overtime to stabilize fuel flow and maintain vehicle control. The fuel cells leaked and had to be redesigned, and the Agena reconfiguration also suffered costly delays. NASA engineers never did get the paraglider to work properly and eventually dropped it from the program in favor of a parachute system like the one used for Mercury. All of these difficulties shot a $350 million program to over $1 billion. The overruns were successfully justified by the space agency, however, as necessities to meet the Apollo landing commitment.

By the end of 1963 most of the difficulties with Gemini had been resolved, albeit at great expense, and the program was ready for flight. Following two unoccupied orbital test flights, the first operational mission took place on 23 March 1965. Mercury astronaut Grissom commanded the mission, with John W. Young (1930–   ), a Naval aviator chosen as an astronaut in 1962, accompanying him. The next mission, flown in June 1965 stayed aloft for four days and astronaut Edward H. White II (1930–1967) performed the first U.S. extravehicular activity (EVA) or spacewalk. Eight more missions followed through November 1966. Despite problems great and small encountered on virtually all of them, the program achieved its goals. Additionally, as a technological learning program Gemini had been a success, with 52 different experiments performed on the 10 missions. The bank of data acquired from Gemini helped to bridge the gap between Mercury and what would be required to complete Apollo within the time constraints directed by the president.

LEARNING ABOUT THE MOON. In addition to the necessity of acquiring the skills to maneuver in space prior to executing the Apollo mandate, NASA had to learn much more about the Moon itself to ensure that its astronauts would survive. They needed to know the composition and geography of Moon, and the nature of the lunar surface. Was it solid enough to support a lander? Was it composed of dust that would swallow up the spacecraft? Would communications systems work on the Moon? Would other factors—geology, geography, radiation, etc.—affect the astronauts? To answer these questions three distinct satellite research programs emerged to study the Moon. The first of these was Project Ranger, which had actually been started in the 1950s, in response to Soviet lunar exploration, but had been a notable failure until the mid-1960s when three probes photographed the lunar surface before crashing into it.

The second project was the Lunar Orbiter, an effort approved in 1960 to place probes in orbit around the Moon. This project, originally not intended to support Apollo, was reconfigured in 1962 and 1963 to further the Kennedy mandate more specifically by mapping the surface. In addition to a powerful camera that could send photographs to Earth tracking stations, it carried three scientific experiments—selnodesy (the lunar equivalent of geodesy), meteoroid detection, and radiation measurement. While the returns from these instruments interested scientists in and of themselves, they were critical to Apollo. NASA launched five Lunar Orbiter satellites between 10 August 1966 and 1 August 1967, all successfully achieving their objectives. At the completion of the third mission, moreover, the Apollo planners announced that they had sufficient data to press on with an astronaut landing and were able to use the last two missions for other activities.

Finally, in 1961 NASA created Project Surveyor to soft-land a satellite on the Moon. A small craft with tripod landing legs, it could take post-landing photographs and perform a variety of other measurements. Surveyor 1 landed on the Moon on 2 June 1966 and transmitted more than 10,000 high-quality photographs of the surface. Although the second mission crash landed, the next flight provided photographs, measurements of the composition and surface-bearing strength of the lunar crust, and readings on the thermal and radar reflectivity of the soil. Although Surveyor 4 failed, by the time of the program's completion in 1968 the remaining three missions had yielded significant scientific data both for Apollo and for the broader lunar science community.

BUILDING SATURN. NASA inherited the effort to develop the Saturn family of boosters used to launch Apollo to the Moon in 1960 when it acquired part of the Army Ballistic Missile Agency under Wernher von Braun. By that time von Braun's engineers were hard at work on the first generation Saturn launch vehicle, a cluster of eight Redstone boosters around a Jupiter fuel tank. Fueled by a combination of liquid oxygen (LOX) and RP-1 (a version of kerosene), the Saturn I could generate a thrust of 205,000 pounds. This group also worked on a second stage, known in its own right as the Centaur, that used a revolutionary fuel mixture of LOX and liquid hydrogen that could generate a greater ratio of thrust to weight. The fuel choice made this second stage a difficult development effort, because the mixture was highly volatile and could not be readily handled. But the stage could produce an additional 90,000 pounds of thrust. The Saturn I was solely a research and development vehicle that would lead toward the accomplishment of Apollo, making ten flights between October 1961 and July 1965. The first four flights tested the first stage, but beginning with the fifth launch the second stage was active and these missions were used to place scientific payloads and Apollo test capsules into orbit.

The next step in Saturn development came with the maturation of the Saturn IB, an upgraded version of the earlier vehicle. With more powerful engines generating 1.6 million pounds of thrust from the first stage, the two-stage combination could place 62,000-pound payloads into Earth orbit. The first flight on 26 February 1966 tested the capability of the booster and the Apollo capsule in a suborbital flight. Two more flights followed in quick succession. Then there was a hiatus of more than a year before the 22 January 1968 launch of a Saturn IB with both an Apollo capsule and a lunar landing module aboard for orbital testing. The only astronaut-occupied flight of the Saturn IB took place between 11 and 22 October 1968 when Walter Schirra, Donn F. Eisele (1930–1987), and R. Walter Cunningham (1932–   ), made 163 orbits testing Apollo equipment.

The largest launch vehicle of this family, the Saturn V, represented the culmination of those earlier booster development and test programs. Standing 363 feet tall, with three stages, this was the vehicle that could take astronauts to the Moon and return them safely to Earth. The first stage generated 7.5 million pounds of thrust from five massive engines developed for the system. These engines, known as the F-1, were some of the most significant engineering accomplishments of the program, requiring the development of new alloys and different construction techniques to withstand the extreme heat and shock of firing. The thunderous sound of the first static test of this stage, taking place at Huntsville, Alabama, on 16 April 1965, brought home to many that the Kennedy goal was within technological grasp. For others, it signaled the magic of technological effort; one engineer even characterized rocket engine technology as a "black art" without rational principles. The second stage presented enormous challenges to NASA engineers and very nearly caused the lunar landing goal to be missed. Consisting of five engines burning LOX and liquid hydrogen, this stage could deliver 1 million pounds of thrust. It was always behind schedule, and required constant attention and additional funding to ensure completion by the deadline for a lunar landing. Both the first and third stages of this Saturn vehicle development program moved forward relatively smoothly. (The third stage was an enlarged and improved version of the IB, and had few developmental complications.)

Despite all of this, the biggest problem with Saturn V lay not with the hardware, but with the clash of philosophies toward development and test. The von Braun "Rocket Team" had made important technological contributions and enjoyed popular acclaim as a result of conservative engineering practices that took minutely incremental approaches toward test and verification. They tested each component of each system individually and then assembled them for a long series of ground tests. Then they would launch each stage individually before assembling the whole system for a long series of flight tests. While this practice ensured thoroughness, it was both costly and time-consuming, and NASA had neither commodity to expend. George E. Mueller (1918–   ), the head of NASA's Office of Manned Space

Flight, disagreed with this approach. Drawing on his experience with the Air Force and aerospace industry, and shadowed by the twin bugaboos of schedule and cost, Mueller advocated what he called the "all-up" concept in which the entire Apollo-Saturn system was tested together in flight without the laborious preliminaries.

A calculated gamble, the first Saturn V test launch took place on 9 November 1967 with the entire Apollo-Saturn combination. A second test followed on 4 April 1968, and even though it was only partially successful because the second stage shut off prematurely and the third stage—needed to start the Apollo payload into lunar trajectory—failed, Mueller declared that the test program had been completed and that the next launch would have astronauts aboard. The gamble paid off. In 17 test and 15 piloted launches, the Saturn booster family scored a 100 percent launch reliability rate.

THE APOLLO SPACECRAFT. Almost with the announcement of the lunar landing commitment in 1961 NASA technicians began a crash program to develop a reasonable configuration for the trip to lunar orbit and back. What they came up with was a three-person command module capable of sustaining human life for two weeks or more in either Earth orbit or in a lunar trajectory; a service module holding oxygen, fuel, maneuvering rockets, fuel cells, and other expendable and life support equipment that could be jettisoned upon reentry to Earth; a retrorocket package attached to the service module for slowing to prepare for reentry; and finally a launch escape system that was discarded upon achieving orbit. The tear-drop shaped command module had two hatches, one on the side for entry and exit of the crew at the beginning and end of the flight and one in the nose with a docking collar for use in moving to and from the lunar landing vehicle.

Work on the Apollo spacecraft stretched from 28 November 1961, when the prime contract for its development was let to North American Aviation, to 22 October 1968 when the last test flight took place. In between there were various efforts to design, build, and test the spacecraft both on the ground and in suborbital and orbital flights. For instance, on 13 May 1964 NASA tested a boilerplate model of the Apollo capsule atop a stubby Little Joe II military booster, and another Apollo capsule actually achieved orbit on 18 September 1964 when it was launched atop a Saturn I. By the end of 1966 NASA leaders declared the Apollo command module ready for human occupancy. The final flight checkout of the spacecraft prior to the lunar flight took place on 11–22 October 1968 with three astronauts.

As these development activities were taking place, tragedy struck the Apollo program. On 27 January 1967, Apollo-Saturn (AS) 204, scheduled to be the first spaceflight with astronauts aboard the capsule, was on the launch pad at Kennedy Space Center, Florida, moving through simulation tests. The three astronauts to fly on this mission—"Gus" Grissom, Edward White, and Roger B. Chaffee (1935–1967)—were aboard running through a mock launch sequence. At 6:31 p.m., after several hours of work, a fire broke out in the spacecraft and the pure oxygen atmosphere intended for the flight helped it burn with intensity. In a flash, flames engulfed the capsule and the astronauts died of asphyxiation. It took the ground crew five minutes to open the hatch. When they did so they found three bodies. Although three other astronauts had been killed before this time—all in plane crashes—these were the first deaths directly attributable to the U.S. space program.

Shock gripped NASA and the nation during the days that followed. James Webb, NASA administrator, told the media at the time, "We've always known that something like this was going to happen soon or later. . . . who would have thought that the first tragedy would be on the ground?" As the nation mourned, Webb went to President Lyndon Johnson and asked that NASA be allowed to handle the accident investigation and direct the recovery from the accident. He promised to be truthful in assessing blame and pledged to assign it to himself and NASA management as appropriate. The day after the fire NASA appointed an eight member investigation board, chaired by longtime NASA official and director of the Langley Research Center, Floyd L. Thompson (1898–1976). It set out to discover the details of the tragedy: what happened, why it happened, could it happen again, what was at fault, and how could NASA recover? The members of the board learned that the fire had been caused by a short circuit in the electrical system that ignited combustible materials in the spacecraft fed by the oxygen atmosphere. They also found that it could have been prevented and called for several modifications to the spacecraft, including a move to a less oxygen-rich environment. Changes to the capsule followed quickly, and within a little more than a year it was ready for flight.

Webb reported these findings to various congressional committees and took a personal grilling at every meeting. The media also attacked him. While the ordeal was personally taxing, whether by happenstance or design Webb deflected much of the backlash over the fire from both NASA as an agency and from the Johnson administration. While he was personally tarred with the disaster, the space agency's image and popular support was largely undamaged. Webb himself never recovered from the stigma of the fire, and when he left NASA in October 1968, even as Apollo was nearing a successful completion, few mourned his departure.

The AS 204 fire also troubled Webb ideologically during the months that followed. He had been a high priest of technocracy ever since coming to NASA in 1961, arguing for the authority of experts, well-organized and led, and with sufficient resources to resolve the "many great economic, social, and political problems" that pressed the nation. He wrote in his book, *Space Age Management*, as late as 1969 that "Our Society has reached a point where its progress and even its survival increasingly depend upon our ability to organize the complex and to do the unusual." He believed he had achieved that model organization for complex accomplishments at NASA. Yet that model structure of exemplary management had failed to anticipate and resolve the shortcomings in the Apollo capsule design and had not taken what seemed in retrospect to be normal precautions to ensure the safety of the crew. The system had broken down. As a result Webb became less trustful of other officials at NASA and gathered more and more decision-making authority to himself. This wore on him during the rest of his time as NASA administrator, and in reality the failure of the technological model for solving problems was an important forecaster of a trend that would be increasingly present in American culture thereafter as technology was blamed for a good many of society's ills. That problem would be particularly present as NASA tried to win political approval of later NASA projects.

THE LUNAR MODULE. If the Saturn launch vehicle and the Apollo spacecraft were difficult technological challenges, the third part of the hardware for the Moon landing, the Lunar Module (LM), represented the most serious problem. Begun a year later than it should have been, the LM was consistently behind schedule and over budget. Much of the problem turned on the demands of devising two separate spacecraft components—one for descent to the Moon and one for ascent back to the command module—that only maneuvered outside an atmosphere. Both engines had to work perfectly or the very real possibility existed that the astronauts would not return home. Guidance, maneuverability, and spacecraft control also caused no end of headaches. The landing structure likewise presented problems; it had to be light and sturdy and shock resistent. An ungainly vehicle emerged which two astronauts could fly while standing. In November 1962 Grumman Aerospace Corp. signed a contract with NASA to produce the LM, and work on it began in earnest. With difficulty the LM was orbited on a Saturn V test launch in January 1968 and judged ready for operation.

TRIPS TO THE MOON. After a piloted orbital mission to test the Apollo equipment on October 1968, on 21 December 1968 Apollo 8 took off atop a Saturn V booster from the Kennedy Space Center with three astronauts aboard—Frank Borman (1928–   ), James A. Lovell, Jr. (1928–   ), and William A. Anders (1933–   )—for a historic mission to orbit the Moon. At first it was planned as a mission to test Apollo hardware in the relatively safe confines of low Earth orbit, but senior engineer George M. Low (1930–1984) of the Manned Spacecraft Center at Houston, Texas, and Samuel C. Phillips, Apollo program manager at NASA Headquarters, pressed for approval to make it a circumlunar flight. (*See Reading No. 15*.) The advantages of this could be important, both in technical and scientific knowledge gained as well as in a public demonstration of what the United States could achieve. So far Apollo had been all promise; now the delivery was about to begin. In the summer of 1968 Low broached the idea to Phillips, who then carried it to the administrator, and in November the agency reconfigured the mission for a lunar trip. After Apollo 8 made one and a half Earth orbits its third stage began a burn to put the spacecraft on a lunar trajectory. As it traveled outward the crew focused a portable television camera on Earth and for the first time humanity saw its home from afar, a tiny, lovely, and fragile "blue marble" hanging in the blackness of space. When it arrived at the Moon on Christmas Eve this image of Earth was even more strongly reinforced when the crew sent images of the planet back while reading the first part of the Bible—"And God created the heavens and the Earth, and the Earth was without form and void"—before sending Christmas greetings to humanity. The next day they fired the boosters for a return flight and splashed down in the Pacific Ocean on 27 December. It was an enormously significant accomplishment coming at a time when American society was in crisis over Vietnam, race relations, urban problems, and a host of other difficulties. And if only for a few moments the nation united as one to focus on this epochal event. Two more Apollo missions occurred before the climax of the program, but they did little more than confirm that the time had come for a lunar landing.

Then came the big event. Apollo 11 lifted off on 16 July 1969, and after confirming that the hardware was working well began the three day trip to the Moon. At 4:18 p.m. EST on 20 July 1969 the LM—with astronauts Neil A. Armstrong (1930–) and Edwin E. Aldrin (1930–)—landed on the lunar surface while Michael Collins (1930–) orbited overhead in the Apollo command module. After checkout, Armstrong set foot on the surface, telling the millions of listeners that it was "one small step for man—one giant leap for mankind." (*See Reading No. 16*.) Aldrin soon followed him out and the two plodded around the landing site in the 1/6 lunar gravity, planted an American flag but omitted claiming the land for the United States as had routinely been done during European exploration of the Americas, collected soil and rock samples, and set up experiments. The next day they launched back to the Apollo capsule orbiting overhead and began the return trip to Earth, splashing down in the Pacific on 24 July.

These flights rekindled the excitement felt in the early 1960s with John Glenn and the Mercury astronauts. An ecstatic reaction circled the globe, as everyone shared in the success of the mission. Ticker tape parades, speaking engagements, public relations events, and a world tour by the astronauts served to create good will both in the United States and abroad. Five more landing missions followed through December 1972, three of them using a lunar rover vehicle to travel in the vicinity of the landing site, but none of them equalled the excitement of Apollo 8 and Apollo 11. Only Apollo 13, launched on 11 April 1970, came close to matching earlier popular interest. But that was only because, 56 hours into the flight, an oxygen tank in the Apollo service module ruptured and damaged several of the power, electrical, and life support systems. People throughout the world watched and waited and hoped as NASA personnel on the ground and the crew, on their way to the Moon and with no way of returning until they went around it, worked together to find a means to return safely home. While NASA engineers quickly determined that air, water, and electricity did not exist in the Apollo capsule sufficient to sustain the three astronauts until they could return to Earth, they found that the LM—a self-contained spacecraft unaffected by the accident—could be used as a "lifeboat" to provide austere life support for the return trip. It was a close-run thing, but the crew returned safely on 17 April 1970. The near disaster served several important purposes for the civil space program—especially prompting reconsideration of the propriety of the whole effort while also solidifying in the popular mind NASA's technological genius. These results were critical (along with several other factors such as budget and national priorities) for the future of the U.S. space program in the post-Apollo era. The Kennedy mandate met, where should space exploration go from there?

CHAPTER 8

NASA IN PARTIAL ECLIPSE

THE LEGACY OF APOLLO. The Apollo program, while an enormous achievement, left a divided legacy for NASA. The sprint to the Moon forced NASA to become largely a single issue agency, and single issue agencies—like single issue political parties—have a difficult time dealing with success. A search for continued meaning always ensues after a goal is achieved. Moreover, the results of Apollo, in contrast to what NASA had wanted, was essentially a technological dead-end for the space program. As former NASA Deputy Administrator Hans Mark (1929–   ) reflected in 1987, "President Kennedy's objective was duly accomplished, but we paid a price: the Apollo program had no logical legacy."

The public's indifference to the accomplishment of Apollo created yet another difficulty for NASA in the post-Apollo era. One child of the sixties recalled that "I once dreamed of men touching the moon. Now I saw it. And I didn't care." This was at least partly because of the age in which the Apollo mandate was accomplished. "We walk safely among the craters of the moon but not in the parks of New York or Chicago or Los Angeles," commented another observer in 1969. The United States in the late 1960s was a nation in crisis, and Apollo seemed to be a relic of an earlier, more simple, more hegemonic time.

The "golden age" of Apollo also created for the agency an expectation that the direction of a major space goal from the president would always bring NASA a broad consensus of support and provide it with the resources and license to dispense them as it saw fit. Something most NASA officials did not understand at the time of the Moon landing in 1969, however, was that this had not been a normal situation and would not be repeated. The Apollo decision was, therefore, an anomaly in the national decision-making process. The dilemma of the "golden age" of Apollo was in large part that NASA came to perceive free-flowing funds, ready political support, and relative autonomy in conducting its activities as normal when such has rarely been the case in U.S. history.

CONSTRICTION. The political consensus that had produced the visionary space exploration program of the early 1960s began to dissipate even before the first Apollo missions were flown. NASA's budget began to decline beginning in 1966 and continued a downward trend until 1975. Diminished priority for the "great national adventure in space" was also felt in other arenas. In 1964, NASA's ten top executives lost their previous special pay status, making the agency unable to compete with the aerospace industry for top management talent. During the latter part of the decade the number of government employees at NASA declined to about two-thirds of its Apollo maximum. The complex organization that managed the massive Apollo program began to contract even before the goal was achieved.

For all of his political acumen, James Webb was ultimately unsuccessful in staving off constriction at the hands of Congress and other parts of the federal government. In part this resulted from major political and social upheavals that reoriented the direction of public policy in the latter part of the decade. The social programs of Johnson's "Great Society" and the Vietnam War consumed large portions of the federal budget and the space program suffered accordingly. Moreover, the nature of warfare in Southeast Asia and the failure of the "Great Society" to solve America's social problems combined to spark a sustained criticism of the national character and meaning that plunged the United States into fundamental change, political turmoil, and activism of all stripes. The space program, viewed as part of the earlier consensus of the American establishment, suffered as well. NASA's fiscal year 1971 budget took a battering, forcing the cancellation of Apollo missions 18 through 20. Moreover, at least by the latter 1960s it was clear to many Americans that the science and technology establishment, of which NASA was a part, did not hold the answers that had been promised. Too often, science and technology were viewed by many as fundamental parts of the problem and not as the solution that they had once seemed.

Although this antitechnological perspective flowed sub rosa

through much of American culture, representatives of the counterculture were especially condemnatory of technical and scientific expertise and its application in modern society. This was partly the result of a belief that the expansion the federal government's role in the lives of individual Americans represented bad government and that the Jeffersonian tradition of local politics and democratic responsiveness to the needs of individuals was being seriously threatened. After all, Jefferson had apostrophized that the best government is the least government. The most radical critics called for, among other things, a return to a more simple existence—the most drastic applications of which were communes of all types—while the more moderate called for resuming control of the political system from the technical experts. The space program, even as Apollo was being completed, lost much of its support in part because of these broader trends in American society.

The constriction was especially difficult during the administration of Richard M. Nixon (1913–   ), who was elected president in 1968. Thomas O. Paine (1921–1992), Nixon's appointee as NASA administrator, described a somber meeting with the president in January 1970 in which Nixon told him a Harris poll reported that 56 percent of Americans believed the costs of the Apollo program were too great and 64 percent believed that $4 billion a year for NASA was too much. Nixon regretted making cuts but said that they were necessary to bring the federal budget into line, pay for expanding entitlement programs, deal with national defense issues, and maintain the infrastructure of government. In such an environment, there was little probability of continuing an expansive space program, and Nixon wanted the NASA budget to shrink to about $3 billion per year. Moreover, he told NASA that Apollo had been successful and asked what more the United States had to prove to the world in the space arena?

Paine put relentless pressure on the president for a greater commitment to NASA's activities, and this led to hard feelings between the White House and senior NASA managers, making all decisions about the space program contentious. An internal White House memorandum dated 8 February 1971 stated that "NASA is—or should be—making a transition from rapid razzle-dazzle growth and glamor to organizational maturity and more stable operations for the long term." Paine, unfortunately, had been unwilling to move in that direction. Officials had concluded: "We need a new Administrator who will turn down NASA's empire-building fervor . . . someone who will work with us rather than against us, and will seek progress toward the President's stated goals, and will shape the program to reflect credit on the President rather than embarrassment." When Paine left NASA on 15 September 1970 Deputy Administrator George Low tried to heal this breech between NASA and the White House. After James C. Fletcher (1924–1992) became NASA administrator in May 1971, he made similar efforts to smooth over these problems and to get on with the definition of a viable post-Apollo spaceflight program. But from that time forward, the U.S. space program had to contend with the reality of tight fiscal constraints.

As support for the civil space program grew softer, the budget and personnel assigned to NASA declined to about half of what they had been during the heyday of Apollo. Faced with deteriorating resources, Administrator Fletcher tried to protect as best he could the technical and scientific core of personnel located at the NASA field centers, the truly essential resources needed to carry out the agency's mission. He designated "roles and missions" for each of the centers, thereby avoiding duplication of effort. This created a particularly difficult environment inside NASA, given the interlocking interests present between installations, contractors, and geographic regions on the one hand and their representatives in Washington on the other. Political infighting became more common as each NASA center struggled for survival. In the end, the NASA centers limped along during the 1970s, losing their best personnel to industry, the military, and universities, but working—usually successfully— to continue a viable space program on a shoestring.

PLANNING THE POST-APOLLO SPACE PROGRAM. As early as January 1964 NASA administrator, James E. Webb, had been asked by the president for a well-developed proposal of future space objectives after Apollo. Webb did not want to respond; instead he tried unsuccessfully to obtain another commitment from the president for a major space effort after the Moon landing. (*See Reading No. 13.*) The consequence was that there were virtually no "new starts" for NASA during this period, and certainly nothing on the order of a new piloted launch system. This was reflected in the NASA budgets of the era. When he left the space agency in October 1968 Webb was embittered by what he saw as a retreat from the aggressive space program of Apollo.

When Richard Nixon took office, he appointed a Space Task Group to study post-Apollo plans and make recommendations. Chartered on 13 February 1969 under the chairmanship of Vice President Spiro T. Agnew (1918–   ), this group met throughout the spring and summer to plot a course for the space program. The politics of this effort was intense. NASA lobbied hard with the group and especially its chair for a far-reaching post-Apollo space program that included development of a space station, a reusable Space Shuttle, a Moon base, and a human expedition to Mars. The NASA position was well reflected in the group's report on 15 September 1969, but Nixon did not act on the group's recommendations. (*See Reading No. 17.*) Instead, he was silent on the future of the U.S. space program for several months. Finally, Nixon issued a 7 March 1970 statement that clearly announced his approach toward dealing with NASA and space exploration, "we must also recognize that many critical problems here on this planet make high priority demands on our attention and our resources." (*See Reading No. 18.*)

Without clear presidential leadership, NASA began to forge ahead on its own, although quietly at first, with what plans it could get approved for a continuation of U.S. spaceflight in the 1970s. Even before it became clear that first Johnson and then Nixon were unwilling to commit the nation to another Apollo-like program, NASA had begun laying the plans for several projects. The first of these was what Webb had called the Apollo Applications Program, since it used some of the technology that had been developed by Project Apollo and since it could be conducted partly under the rubric of the larger lunar mission. At the heart of this effort was a relatively small orbital space platform, called Skylab, that could be tended by astronauts. It would be, NASA officials hoped, the precursor of a real space station. The second effort was a diplomatic mission, later named the Apollo-Soyuz Test Project, in which Americans and Soviets rendezvoused and docked in space. A third project involved the soft landing of two probes on Mars, Project Viking. Another major goal in the immediate post-Apollo era was sending satellite probes to the outer planets of the solar system, sometimes called the "Grand Tour." Finally, NASA became involved in Earth resource mapping from space.

SKYLAB. America's first experimental space station, this program originated in the 1960s to prove that humans could live and work in space for extended periods and to expand knowledge of solar astronomy beyond what could be achieved from Earth-based observations. It made extensive use of Saturn and Apollo equipment, an idea that had been germinating within NASA since 1963, by using a reconfigured and habitable third stage of the Saturn V rocket as the basic component of the orbital station. NASA engineers developed two concepts for this space platform. The first was dubbed the "wet" approach, in which a Saturn upper stage that had been used to achieve orbit would be refurbished for habitation by astronauts in space. This concept proved too risky and difficult, so NASA opted for a "dry" concept in which the stage was completely outfitted as an orbital workshop before launch. Carried out on a shoestring for a human spaceflight program, in large measure because of the use of equipment developed and built with Project Apollo funding, the direct Skylab expenditure cost less than $3 billion.

The 100-ton orbital workshop was launched into orbit on 14 May 1973, the last use of the giant Saturn V launch vehicle. Almost immediately, technical problems developed due to vibrations during lift-off. Sixty-three seconds after launch, the meteoroid shield—designed also to shade Skylab's workshop from the Sun's rays—ripped off, taking with it one of the spacecraft's two solar panels, and another piece wrapped around the other panel that kept it from properly deploying. In spite of this, the space station achieved a near-circular orbit at the desired altitude of 270 miles. NASA's mission control personnel maneuvered Skylab so that its Apollo Telescope Mount (ATM) solar panels faced the Sun to provide as much electricity as possible, but because of the loss of the meteoroid shield, this positioning caused workshop temperatures to rise to 126 degrees Fahrenheit.

While NASA technicians worked on a solution to the problem

Skylab 2, the first mission with astronauts, was postponed. In an intensive ten-day period, NASA developed procedures and trained the crew to make the workshop habitable. At the same time, engineers "rolled" Skylab to lower the temperature of the workshop. Finally on 25 May 1973 astronauts Charles Conrad, Jr. (1930–   ), Paul J. Weitz (1932–   ), and Joseph P. Kerwin (1932–   ), lifted off from Kennedy Space Center in an Apollo capsule atop a Saturn IB and rendezvoused with the orbital workshop. After substantial repairs requiring extravehicular activity (EVA), including deployment of a parasol sunshade that cooled the inside temperatures to 75 degrees Fahrenheit, by 4 June the workshop was in full operation. In orbit the crew conducted solar astronomy and Earth resources experiments, medical studies, and five student experiments. This crew made 404 orbits and carried out experiments for 392 hours, in the process making three EVAs totalling six hours and 20 minutes. The first group of astronauts returned to Earth on 22 June 1973, and two other Skylab missions followed.

NASA was delighted with the scientific return from the Skylab program despite its early and reoccurring mechanical difficulties. A total of three three-person crews occupied the Skylab workshop for a total of 171 days and 13 hours. It was the site of nearly 300 scientific and technical experiments. In Skylab, both the total hours in space and the total hours spent in performance of EVA under microgravity conditions exceeded the combined totals of all of the world's previous space flights up to that time.

Following the final occupied phase of the Skylab mission, ground controllers performed some engineering tests of certain Skylab systems—tests that ground personnel were reluctant to do while astronauts were aboard—positioned Skylab into a stable attitude and shut down its systems. It was expected that Skylab would remain in orbit eight to ten years, by which time NASA might be able to reactivate it. In the fall of 1977, however, agency officials determined that Skylab had entered a rapidly decaying orbit—resulting from greater than predicted solar activity—and that it would reenter the Earth's atmosphere within two years. They steered the orbital workshop as best they could so that debris from reentry would fall over oceans and unpopulated areas of the planet. On 11 July 1979, Skylab finally impacted the Earth's surface. The debris dispersion area stretched from the Southeastern Indian Ocean across a sparsely populated section of Western Australia. NASA and the U.S. space program took criticism for this development, ranging from the sale of hardhats as "Skylab Survival Kits" to serious questions about the propriety of spaceflight altogether if people were likely to be killed by falling objects. In reality, while NASA took sufficient precautions so that no one was injured, its leaders had learned that the agency could never again allow a situation in which large chunks of orbital debris had a chance of reaching the Earth's surface.

APOLLO-SOYUZ TEST PROJECT. This was the first international human spaceflight, taking place at the height of the detente between the United States and the Soviet Union during the mid-1970s. It was specifically designed to test the compatibility of rendezvous and docking systems for American and Soviet spacecraft, and to open the way for international space rescue as well as future joint manned flights. To carry out this mission existing American Apollo and Soviet Soyuz spacecraft were used. The Apollo spacecraft was nearly identical to the one that orbited the Moon and later carried astronauts to Skylab, while the Soyuz craft was the primary Soviet vehicle used for cosmonaut flight since its introduction in 1967. A universal docking module was designed and constructed by NASA to serve as an airlock and transfer corridor between the two craft.

The actual flight took place between 15 and 24 July 1975 when astronauts Thomas P. Stafford (1930–   ), Vance D. Brand (1931–   ), and Donald K. Slayton took off from Kennedy Space Center to meet the already orbiting Soyuz spacecraft. Some 45 hours later the two craft rendezvoused and docked, and then Apollo and Soyuz crews conducted a variety of experiments over a two-day period. After separation, the Apollo vehicle remained in space an additional six days while Soyuz returned to Earth approximately 43 hours after separation. The flight was more a symbol of the lessening of tensions between the two superpowers than a significant scientific endeavor, taking 180 degrees the competition for international prestige that had fueled much of the space activities of both nations since the late 1950s.

VIKING. Project Viking was the culmination of a series of missions to explore the planet Mars that had begun in 1964 with Mariner 4, and continued with the Mariner 6 and Mariner 7 flybys in 1969 and the Mariner 9 orbital mission in 1971 and 1972. (*See Reading No. 14.*) After losing a more ambitious and expensive program in the late 1960s, NASA came forward with a somewhat more modest $1 billion budget for the Viking expedition to the Red Planet. This purchased tandem spacecraft designed to orbit Mars and to land and operate on the planet's surface. Two identical spacecraft, each consisting of a lander and an orbiter, were built. Launched on 20 August 1975 from the Kennedy Space Center, Viking 1 spent nearly a year cruising to Mars, placed an orbiter in operation around the planet, and landed on 20 July 1976 on the Chryse Planitia (Golden Plains). Viking 2 was launched on 9 September 1975 and landed on 3 September 1976. The Viking project's primary mission ended on 15 November 1976, 11 days before Mars' superior conjunction (its passage behind the Sun), although the Viking spacecraft continued to operate for six years after first reaching Mars. Its last transmission reached Earth on 11 November 1982. Controllers at NASA's Jet Propulsion Laboratory tried unsuccessfully for another six and one-half months to regain contact with the lander, but finally closed down the overall mission on 21 May 1983.

With a single exception—the seismic instruments—the scientific return from the expedition was spectacular. Unfortunately, the seismometer on Viking 1 did not work after landing, and the seismometer on Viking 2 detected only one event that may have been seismic. On the other hand, the two landers continuously monitored weather at the landing sites and found both exciting cyclical variations and an exceptionally harsh climate. Atmospheric temperatures at the more southern Viking 1 landing site, for instance, were only as high as $+7$ degrees Fahrenheit at midday, but the predawn summer temperature was -107 degree Fahrenheit. And the lowest predawn temperature was -184 degrees Fahrenheit, about the frost point of carbon dioxide. The project also observed the Martian winds, finding that they generally blew more slowly than expected.

One of the important scientific activities of this project was the attempt to determine whether there was life on Mars, since the planet had long been thought of as having sufficient similarity to the Earth that life might exist there. While the three biology experiments discovered unexpected and enigmatic chemical activity in the Martian soil, they provided no clear evidence for the presence of living microorganisms in soil near the landing sites. According to mission biologists, Mars was self-sterilizing. They concluded that the combination of solar ultraviolet radiation that saturates the surface, the extreme dryness of the soil, and the oxidizing nature of the soil chemistry had prevented the formation of living organisms in the Martian soil. The question of life on Mars at some time in the distant past, however, remains open.

THE GRAND TOUR. During the latter 1960s JPL scientists and others discovered that once every 176 years both the Earth and all the giant planets of the solar system gather on one side of the Sun. This geometric line-up made possible close-up observation of all the planets in the outer solar system (with the exception of Pluto) in a single flight, the "Grand Tour." The flyby of each planet would bend the spacecraft's flight path and increase its velocity enough to deliver it to the next destination. This would occur through a complicated process known as "gravity assist," something like a slingshot effect, whereby the flight time to Neptune could be reduced from 30 to 12 years. Such a configuration was due to occur in the late 1970s, and it led to one of the most significant space probe efforts undertaken by the United States.

To prepare the way for the "Grand Tour," NASA conceived Pioneer 10 and Pioneer 11 as outer solar system probes. Both were small, nuclear-powered, spin-stabilized spacecraft that Atlas-Centaur launched. The first of these was launched on 3 March 1972, traveled outward to Jupiter, and in May 1991 was about 52 Astronautical Units (AU), roughly twice the distance from Jupiter to the Sun, and still transmitting data. In 1973, NASA launched Pioneer 11, providing scientists with their closest view of Jupiter, from 26,600 miles above the cloud tops in December 1974. The close approach and the spacecraft's speed of 107,373 mph, by far the fastest ever reached by an object from Earth, hurtled Pioneer 11 outward.

Meantime, NASA technicians prepared to launch what was called Project Voyager. While the four-planet mission was known to be possible, it was quickly deemed too expensive to build a spacecraft that could go the distance, carry the instruments needed, and last long enough to accomplish such an extended mission. Thus, the two Voyager spacecraft were funded to conduct intensive flyby studies only of Jupiter and Saturn, in effect repeating on a more elaborate scale the flights of the two Pioneers. Even so, the spacecraft builders designed as much longevity into the two Voyagers as possible with the $865 million budget available. NASA launched these from Cape Canaveral, Florida: Voyager 2 lifting off on 20 August 1977 and Voyager 1 entering space on a faster, shorter trajectory on 5 September 1977. Both spacecraft were delivered to space aboard Titan-Centaur expendable rockets.

As the mission progressed, with the successful achievement of all its objectives at Jupiter and Saturn in December 1980, additional flybys of the two outermost giant planets, Uranus and Neptune, proved possible—and irresistible—to mission scientists and engineers at the Jet Propulsion Laboratory in Pasadena, California. Accordingly, as the spacecraft flew across the solar system, remote-control reprogramming was used to reprogram the Voyagers for the greater mission. Eventually, between them, Voyager 1 and Voyager 2 explored all the giant outer planets, 48 of their moons, and the unique systems of rings and magnetic fields those planets possess.

The two spacecraft returned to Earth information that has revolutionized the science of planetary astronomy, helping to resolve some key questions while raising intriguing new ones about the origin and evolution of the planets in this solar system. The two Voyagers took well over 100,000 images of the outer planets, rings, and satellites, as well as millions of magnetic, chemical spectra, and radiation measurements. They discovered rings around Jupiter, volcanoes on Io, shepherding satellites in Saturn's rings, new moons around Uranus and Neptune, and geysers on Triton. The last imaging sequence was Voyager 1's portrait of most of the solar system, showing Earth and six other planets as sparks in a dark sky lit by a single bright star, the Sun. The Voyagers are expected to return scientific data until about

2010 since communications will be maintained until their nuclear power sources can no longer supply enough electrical
energy to power critical subsystems.

LANDSAT. In addition to scientific and human space flight
activities, NASA began in the 1970s to build and launch Earth
resource mapping satellites, the first of which was the Landsat
series. Landsat 1, launched on 23 July 1972 as the Earth
Resources Technology Satellite (ERTS) and later renamed,
changed the way in which Americans looked at the planet. It
provided data on vegetation, insect infestations, crop growth,
and associated land-use information. Two more Landsat vehicles were launched in January 1975 and March 1978, performed
their missions and exited service in the 1980s. Landsat 4,
launched 16 July 1982, and Landsat 5, launched 1 March 1984,
were "second generation" spacecraft, with greater capabilities
to produce more detailed land-use data. The system enhanced
the ability to develop a world-wide crop forecasting system.
Moreover, Landsat imagery has been used to devise a strategy
for deploying equipment to contain oil spills, to aid navigation,
to monitor pollution, to assist in water management, to site new
power plants and pipelines, and to aid in agricultural development.

ASSESSMENT. At the conclusion of Project Apollo, NASA's
overall budget shrank to about $3 billion per year as the press of
other national priorities emerged. As a result, the U.S. space
program entered a period of relative decline. The technological
capability that had been built up to accomplish the lunar landing
within the time constraints established by President Kennedy
began first to atrophy and then to be dismantled in the austere
1970s. This fact weighed heavily on NASA leaders who regret-
ted this political decision. They urged the continuation of an
aggressive space program that would lead to human exploration
of the solar system, but without success. Instead, they set
priorities for the use of a more limited budget, rescoped their
goals, and executed an impressive array of space missions
during the 1970s, many with spectacular scientific results. (*See
Reading No. 22.*) What truly suffered during the era was the
human spaceflight effort, with no astronauts in space between

the completion of the ASTP project in 1975 and the first orbital flight of the Space Shuttle in 1981. Because of the activities of the space program under difficult fiscal conditions Representative George E. Brown, Jr., (1920–   ) (D-CA), Chair of the House committee overseeing NASA, remarked in 1992—when similar spartan budgets appeared in the offing—"It is also important to recall that some of our proudest achievements in the space program have been accomplished within a stagnant, no growth budget. The development of the Landsat program, the Viking lander, Voyagers I and II, Pioneer Venus, and even the Space Shuttle were all carried out during the 1970s when the NASA budget was flat. It would be wise to review how we set priorities and managed programs during this productive time."

CHAPTER 9

ROUTINE ACCESS TO SPACE?

In the austere 1970s NASA undertook what proved in retrospect to be both a technological wonder and nightmare for the agency. As early as the mid-1960s, NASA began the process of developing a reusable Space Shuttle that would be able to travel back and forth between Earth and space more routinely and economically than had ever been the case before. But the Shuttle became the major program of the agency in the 1970s and 1980s, and used nearly a third of the agency's budget in the 1990s. The bobs and weaves, ins and outs, ups and downs of the Space Shuttle effort in the 1970s and 1980s have contributed both high drama and low comedy to American perceptions of the civil space program.

THE SPACE SHUTTLE DECISION. The Space Shuttle had been conceived early in NASA's history as an integral part of a much larger program to provide logistics support to a space station. The goal of the new vehicle was simple, to provide "routine access to space" at an economical cost. Some NASA officials compared the methods of launching into orbit used on Project Apollo to operating a railroad and throwing away the locomotive after every trip. A reusable Shuttle, they claimed, would make the trip much more cost effective. Studies NASA conducted in the mid-1960s found that reusable space technology was within reasonable grasp, more evolutionary than revolutionary, and that a hefty investment of research and development funds could yield a substantial reduction in operations costs. Flying 30 or more times a year, such a system would be an economical alternative to the use of large "throw away" launchers like the Saturn.

The goal of efficient operations in a heavy-lift booster—especially with the decision for budgetary reasons to terminate the Saturn V booster production line in mid-1968 after the completion of 15 launch vehicles—prompted NASA's commitment to the Shuttle as a continuation vehicle for human space flight. Once it was underway, NASA leaders believed, they could also move forward with a space station, which the Shuttle could both place in orbit and support logistically. In addition, and this was pure serendipity from the NASA perspective, because of the Shuttle's size and versatility a portion of its payload bay could be used to haul scientific and applications satellites of all types into orbit for all users. The Shuttle was to be, essentially, the achievement of one-size-fits-all, in this instance the vehicle providing all orbital services required by users. This type of standardization has long been an important part of American mass production, the Model-T automobile and the F-111 fighter-bomber being examples of how it was supposed to work.

NASA had hoped in the fall of 1969 that Nixon would embrace its aggressive plan not only to build a space station and a Shuttle, but also to mount an eventual Mars mission. The president did not agree, forcing the space agency to back away from its overall goals and accept something less. That "something less" became the Space Shuttle. But even in this case there were difficulties in gaining presidential approval. That was largely because of the projected $10–$15 billion investment in research and development (R&D) that could not be recouped until the Shuttle began to fly as a routine system so as to reduce overall launch costs. As a result of this situation, the Nixon administration began to put pressure on NASA to reduce the project's overall R&D costs to a level of about $5 billion before it would endorse the program.

To meet this goal NASA studied a variety of Shuttle configurations. Some of them were quite exotic, but the underlying assumption in 1969 was that the Shuttle would be a two-stage, fully-reusable system with both stages piloted and capable of landing on a runway like conventional aircraft. Launched like a rocket, the two stages would separate near the edge of the atmosphere, with the first stage, about the size of a Boeing 747, returning to Earth. The second stage—about the size of a Boeing 707—would fly on under its own power into orbit, perform its mission, and then return to Earth. NASA got the cost of this system down to an estimated $10 billion, but this was still too high for approval from Nixon. The agency went back to the drawing board, and during 1970 and 1971 it studied additional configurations. The Space Shuttle that finally emerged was a partially reusable vehicle that could be launched atop a new, partially expendable booster system, while a second stage flew into orbit, performed its mission, and then returned to earth. This system could be developed, NASA estimated, for the bargain price of $5.5 billion.

NASA Administrator James Fletcher worked hard to sell this Shuttle concept during the first half of 1971 without any apparent success. But late in the summer he found a receptive audience in Casper Weinberger (1917–   ), Deputy Director of the Office of Management and Budget (OMB), who carried NASA's case to the president. Weinberger wrote a 12 August 1971 memorandum to Nixon, arguing that "there is real merit to the future of NASA, and to its proposed programs." The memo suggested that further cuts to NASA's budget "would be confirming in some respects, a belief that I fear is gaining credence at home and abroad: That our best years are behind us, that we are turning inward, reducing our defense commitments, and voluntarily starting to give up our super-power status, and our desire to maintain world superiority." Weinberger added that "America should be able to afford something besides increased welfare, programs to repair our cities, or Appalachian relief and the like." In a handwritten scrawl on Weinberger's memo, Nixon wrote, "I agree with Cap." (*See Reading No. 19.*)

Fletcher did not know of this exchange, and in the summer and fall of 1971 he led an often heated debate with administration bureaucrats to gain approval of the Shuttle program. In a 22 November 1971 memo to the president he based the justification of the Shuttle on these reasons:

1.   The U.S. cannot forego manned space flight.

2.   The space shuttle is the only meaningful new manned space flight program that can be accomplished on a modest budget.

3.   The space shuttle is a necessary next step for the practical use of space . . .

4.   The cost and complexity of today's shuttle is *one-half* of what it was six months ago.

5.   Starting the shuttle now will have a significant positive

effect on aerospace employment. Not starting would be a serious blow to both the morale and health of the [U.S.] Aerospace Industry.

Toward the end of 1971 Fletcher learned that Nixon had decided in principle to go ahead with the Shuttle project, but that some additional decisions over size and cost had yet to be made. The perspective that Weinberger had put forward in August, Fletcher's arguments in his November memorandum, and the desire to start a new aerospace program that would avoid unemployment in critical states in the 1972 election year ultimately proved decisive. On the evening of 4 January 1972 Fletcher flew from Washington, D.C., to San Clemente, California, where President Nixon was on a working vacation. The next day he met briefly with the president to discuss the future of the space program and then issued a statement to the media announcing the decision to "proceed at once with the development of an entirely new type of space transportation system designed to help transform the space frontier of the 1970s into familiar territory, easily accessible for human endeavor in the 1980s and '90s." (*See Reading No. 21.*) The Shuttle became the largest, most expensive, and highly visible project undertaken by NASA after its first decade, and it has continued to be a central component in the U.S. space program in the 1990s.

DEVELOPING THE SPACE SHUTTLE. The Space Shuttle that emerged in the early 1970s consisted of three primary elements: a delta-winged orbiter spacecraft with a large crew compartment, a 15 by 60 feet cargo bay, and three main engines; two Solid Rocket Boosters (SRB); and an external fuel tank housing the liquid hydrogen and oxidizer burned in the main engines. The orbiter and the two solid rocket boosters were reusable. The Shuttle was designed to transport approximately 45,000 tons of cargo into near-Earth orbit, 100 to 217 nautical miles (115 to 250 statute miles) above the Earth. It could also accommodate a flight crew of up to 10 persons, although a crew of 7 would be more common, for a basic space mission of seven days. During a return to Earth, the orbiter was designed so that it had a cross-range maneuvering capability of 1,100 nautical miles (1,265 statute miles) to meet requirements for liftoff and landing at the same location after only one orbit. This capability satisfied the Department of Defense's need for the Shuttle to place in orbit and retrieve reconnaissance satellites.

NASA began the process of developing the Shuttle on 31 March 1972 when it selected Rockwell International to design and develop the main engines. Contracts followed to Martin Marietta for the external fuel tank on 16 August 1973 and to Morton Thiokol for the solid rocket boosters in June 1972. The next month, 26 July 1972, NASA selected Rockwell to design and build a test orbiter and four operational vehicles.

There were several challenges to be met in building this entirely new type of space vehicle. Perhaps the most important design issue, after the orbiter's configuration, concerned the boosters to be developed and whether they should burn liquid or solid fuel. Also important was the development of the unique reentry method of the Shuttle orbiter. A question arose over how best to pass through the ionosphere—with a high angle of attack that brought the orbiter through it quickly and heated the skin to extremely high temperatures but for a short period of time, or by using a more gradual blunt-body approach that dissipated heating from friction like that of earlier capsules. NASA eventually decided on the blunt-body approach and thus had to develop a special ceramic tile to be placed on the underside and nose of the orbiter to withstand the reentry heat. Because of these issues, as well as political and management questions, the Shuttle development program bogged down seriously in the mid-1970s, prompting its redefinition and refinancing and a delay of its first operational flight from 1979 to 1981.

The first orbiter, *Enterprise* (OV-101)—named for the spacecraft made famous in the "Star Trek" television series after a promotional campaign by "trekkers" such as had never been seen before in space program history—was rolled out of the Rockwell contractor plant on 17 September 1976. In January 1977 the *Enterprise* was taken overland to NASA's Dryden Flight Research Facility in southern California and made its first flight atop its Boeing 747 test platform on 18 February. Such "captive" tests continued through most of the summer of 1977, but on 12 August the first free-flight took place. Some difficulties did materialize in this test program. On the fifth and last free flight on 26 October 1977 the *Enterprise* encountered control problems at touchdown. While trying to slow the spacecraft for landing the pilot experienced a left roll, corrected for it, and touched down too hard. The Shuttle bounced once and eventually settled down to a longer landing than expected. This "Pilot Induced Oscillation," as it was called, was occasioned by the pilot taking over from an automated system too late and not allowing himself sufficient time to get the "feel" of the craft. It was, fortunately, self-correcting when the pilot relaxed the controls, and the positive result led to a decision to take the *Enterprise* on to the Marshall Space Flight Center in Huntsville, Alabama, for a series of ground vibration tests. Several other test elements—engines and such—were completed during the latter part of the 1970s, each directing the program toward an orbital flight in 1981. This would occur in another orbiter since the *Enterprise* had been designed only for test purposes and was later retired to the Smithsonian Institution.

FIRST FLIGHTS. There was tremendous excitement when *Columbia*, the first orbiter that could be flown in space, took off from Cape Canaveral, Florida, on 12 April 1981, six years after the last American astronaut had returned from space. The vehicle was piloted by astronauts John W. Young (1930–) and Robert L. Crippen (1937–). At launch, the orbiter's three liquid-fueled engines—drawing propellants from the external tank—and the two solid rocket boosters generated approximately seven million pounds of thrust. After about two minutes, at an altitude of around 31 miles, the two boosters were spent and separated from the external tank. Waiting ships recovered them for eventual refurbishment and reuse on later missions. The spacecraft's main engines continued to fire for about eight minutes more before shutting down just as the Shuttle entered orbit. As they did so the external tank separated from the orbiter and followed a ballistic trajectory back to the ocean but was not recovered. The orbiter reached a velocity on orbit of approximately 17,322 statute miles per hour, making a circle of the globe in less than two hours. After two days in space, where the crew tested *Columbia*, excitement permeated the nation once again as it landed like an aircraft at Edwards Air Force Base, California. The first flight had been a success, and both NASA and the media ballyhooed the beginning of a new age in spaceflight, one in which there would be inexpensive and routine access to space for many people and payloads. Speculations abounded that within a few years Shuttle flights would take off and land as predictably as airplanes and that commercial tickets would be sold for regularly scheduled "spaceline" flights.

NASA went on to build three additional reusable orbiter spacecraft in addition to *Columbia*. All were named after famous exploration sailing ships. *Columbia* (OV-102) commemorated one of the first U.S. Navy ships to circumnavigate the globe in 1836. *Challenger* (OV-099) was named for the Navy ship that made a prolonged exploration of the Atlantic and Pacific Oceans between 1872 and 1876. *Discovery* (OV-103) was named for two ships, the vessel in which Henry Hudson searched in 1610–1611 for a Northwest Passage between the Atlantic and Pacific Oceans and instead discovered Hudson Bay and the ship in which Captain Cook visited the Hawaiian Islands and explored southern Alaska and western Canada. Finally, *Atlantis* (OV-104) was named after a two-masted ketch operated for the Woods Hole Oceanographic Institute from 1930 to 1966 that traveled more than half a million miles in oceanic research.

Several of the early Shuttle missions were memorable. In spite of some difficulties (*See Reading No. 20.*) space science got something of a boost from the Shuttle since it could take into orbit large numbers of experiments. The boon to atmospheric and space physics, high energy astrophysics, infrared astronomy, optical and ultraviolet astronomy, life sciences, and materials research was often held up as important in the program. One of the "successes" of the Shuttle science program was the so-called "Get Away Special," a set of containers available on all Shuttle missions that allowed both professional and nonprofessional experimenters to gain access to space for small-scale and relatively untended experiments. One of the most significant aspects of space science aboard the Shuttle was the development of "Spacelab," a sophisticated laboratory built by the European Space Agency, that fit into the cargo bay. First flying in 1983, this miniature laboratory had made four flights on the Shuttle before 1986. The Shuttle also demonstrated something of its promised benefits in April 1984 when its astronauts retrieved, repaired, and reorbited the ailing Solar Max satellite. Even so, some scientists questioned the use of the Shuttle for scientific activities and suggested that the developmental costs could more usefully have been applied to expendable systems and robotic probes that promised higher scientific returns on investments.

The Shuttle also greatly expanded the opportunity for human spaceflight. By the time of its tenth anniversary in 1991 it had flown 204 people, some of them more than once. Among other notable developments, in June 1983 Dr. Sally K. Ride (1951–   ), a NASA scientist-astronaut, became the first American woman to fly in space. The Shuttle era also saw flights of people who were not truly astronauts. Senator Jake Garn (1932–   ) (R-UT) and Representative Bill Nelson (1942–   ) (D-FL) both left Congress long enough to fly on the Shuttle in 1985 and 1986. After his flight, Nelson, chair of the House Space Science and Applications Subcommittee, offered this assessment of the space program: "If America ever abandoned her space ventures, then we would die as a nation, becoming second-rate in our own eyes, as well as in the eyes of the world. . . . Our prime reason for commitment can be summed up as follows . . . space is our next frontier." In addition, the first teacher in space, Christa McAuliffe (1948–1986), died in the *Challenger* accident in January 1986, and astronauts from other nations flew aboard the Shuttle during other missions. While this expansion of access to spaceflight was positively received in most instances, some critics accused NASA of pandering to Congress and other constituencies for support by offering such perquisites to a carefully selected few.

In spite of the high hopes that had attended the first launch of *Columbia* in 1981, the Shuttle program provided neither inexpensive nor routine access to space. By January 1986, there had been only 24 Shuttle flights, although in the 1970s NASA had projected more flights than that for every year. While the system was reusable, its complexity, coupled with the ever-present rigors of flying in an aerospace environment, meant that the turnaround time between flights was several months instead of several days. In addition, missions were delayed for all manner of problems associated with ensuring the safety and performance of such a complex system. Since the flight schedule did not meet expectations, and since it took thousands of work hours and expensive parts to keep the system performing satisfactorily, observers began to criticize NASA for failing to meet the cost-effectiveness expectations that had been used to gain the approval of the Shuttle program 10 years earlier. Critical analyses agreed that the Shuttle had proven to be neither cheap nor reliable, both primary selling points, and that NASA should never have used those arguments in building a political consensus for the program. In some respects, therefore, there was some agreement by 1985 that the effort had been both a triumph and a tragedy. The program had been an engagingly ambitious one that had developed an exceptionally sophisticated vehicle, one that no other nation on Earth could have built at the time. As such it had been an enormously successful program. At the same time, the Shuttle was essentially a continuation of space spectaculars, à la Apollo, and its much-touted capabilities had not been realized. It made far fewer flights and conducted far fewer scientific experiments than supporters had publicly predicted.

THE *CHALLENGER* ACCIDENT. All of these criticisms reached crescendo proportions following the tragic loss of *Challenger* during a launch on 28 January 1986. Although it was not the entire reason, the pressure to get the Shuttle schedule more in line with earlier projections throughout 1985 prompted NASA workers to accept operational procedures that fostered short-cuts and increased the opportunity for disaster. It came 73 seconds into the flight, about 11:40 a.m. E.S.T., as a result of a leak in one of two solid rocket boosters that detonated the main liquid fuel tank. Seven astronauts—Francis R. Scobee (1939–1986), Michael J. Smith (1945–1986), Judith A. Resnik (1949–1986), Ronald E. McNair (1950–1986), Ellison S. Onizuka (1946–1986), Gregory B. Jarvis (1944–1986), and Christa McAuliffe—died in this accident, the worst in the history of space flight. The accident, traumatic for the American people even under the best of situations, was made that much worse because the *Challenger's* crewmembers represented a cross-section of the American population in terms of race, gender, geography, background, and religion. The explosion became one of the most significant events of the 1980s, as billions around the world saw the accident on television and empathized with any one or more of the crewmembers killed.

Several investigations followed the accident, the most important being the presidentially mandated blue ribbon commission chaired by William P. Rogers (1913–   ). (*See Reading No. 24.*) It found that the *Challenger* accident resulted from a poor engineering decision, an O-ring used to seal joints in the Solid Rocket Booster that was susceptible to failure at low temperatures, introduced innocently enough years earlier. Rogers kept the commission's analysis on that technical level, and documented the problems in exceptional detail. The commission, after some prodding by Nobel Prize-winning scientist Richard P. Feynman (1918–1988), did a credible if not unimpeachable job of grappling with the technologically difficult issues associated with the *Challenger* accident.

The Rogers Commission also criticized the communication system inside NASA, finding that the potential for O-ring failure had been understood by NASA engineers prior to the launch of *Challenger* and that the accident could have been avoided. With the benefit of hindsight, the Commission took its findings about the accident as a starting point and read history backwards on a trail that led directly to faulty communications of known dangers concerning the O-rings of the Solid Rocket Booster. The Commission quickly decided that the joint the O-rings were supposed to seal was hazardous and that project engineers knew it. It also charged that Marshall Space Flight Center personnel failed to advise their superiors of that hazard because of a poor communication system. Some journalists went even further and explicitly made the case that NASA leaders knew of the O-ring problem but still pressed operations officials to launch so that President Ronald Reagan (1911–   ) could mention the flight in his State of the Union Address that evening.

In spite of these conclusions, there is evidence to indicate that the *Challenger* accident had more to do with NASA's organizational patterns and technological decisions that had made sense at the time they were enacted—mostly in the austere period of the early 1970s—but that in retrospect turned out to be faulty than with the communication issue dwelt upon by the Rogers Commission. Detailed analysis of both documentary evidence and testimony showed that NASA personnel involved in the O-ring question were convinced that the joints were safe, and that there were numerous other problems— especially with the Shuttle main engines—that consumed most of their attention. Most importantly, there had been little engineering data at the time of the accident to support a correlation between O-ring anomalies and low temperatures. On the other hand NASA engineers intuitively knew, were working on correcting, and communicated to others their belief that anomalies had occurred during previous cold weather launches. The fact that the seals had always done their jobs before contributed to a sense that they would not cause a major accident. The catastrophic failure that occurred was a total shock to the NASA staff, made all the more painful by their perception that the Commission had used the agency as a scapegoat to deflect blame away from political leaders in Washington. In a sense, therefore, both the *Challenger* accident, and its official investigation, said less about the overall space effort, the Shuttle program, and the O-rings that allowed the explosion of the spacecraft, than about the political system that produced them.

RETURN TO FLIGHT. With the *Challenger* accident, the Shuttle program went into a two-year hiatus while NASA worked to redesign the solid rocket boosters and revamp its management structure. James C. Fletcher, the NASA Administrator between 1971 and 1977, was brought back to head the space agency a second time with the specific task of overhauling it. He brought former Shuttle astronaut Richard H. Truly (1937–   ) to take charge of the Shuttle program; other personnel changes followed quickly thereafter. NASA reinvested heavily in its safety and reliability programs, made organizational changes to improve efficiency, and restructured its management system. The space agency also built a replacement orbiter for *Challenger*, *Endeavour* (OV-105), named for the first ship Captain Cook commanded on his voyage of discovery in the Pacific in 1768. Most important, NASA engineers completely reworked the components of the Shuttle to enhance its safety and added an egress method for the astronauts. A critical decision resulting from the accident and its aftermath—during which the nation experienced a reduction in capability to launch satellites—was to expand greatly the use of expendable launch vehicles.

The Space Shuttle finally returned to flight without further incident on 29 September 1988. Through April 1993 NASA had launched an additional 30 Shuttle missions without an accident. Each undertook scientific and technological experiments ranging from the deployment of important space probes like the Magellan Venus radar mapper in 1989 and the Hubble Space Telescope in 1990, through the continued flight of "Spacelab" in 1991 through a dramatic three-person EVA in 1992 to retrieve a satellite and bring it back to Earth for repair, to an exciting repair mission for the Hubble Space Telescope in December 1993. Through all of these activities, a good deal of realism about what the Shuttle could and could not do began to emerge in the latter 1980s.

ASSESSMENT. By the early 1990s NASA's Space Shuttle program had reached a crossroads. It was no longer simply a test program in which Americans were obtaining exceptional knowledge about the harsh environment of space or scientific details about the universe or any of the other myriad of possibilities that fueled much of the excitement for the Shuttle in the first place. Moreover, the claims to "routine access to space" and inexpensive operations had not been realized. It had never flown more than nine missions per year, and that many only in 1985, and its support structure gobbled up almost one-third of NASA's nearly $15 billion budget. The Space Shuttle, while a successful program in many ways—especially as a demonstration of technological ability and an important stop on the route toward a truly viable spaceplane—was a disappointment in many others.

Accordingly, on 24 July 1991 Vice President Dan Quayle, as chair of the National Aeronautics and Space Council, issued a report that called for NASA to employ a mixed fleet of expendable launch vehicles to complement the Space Shuttle. The intent of the plan was to reduce dependence on the Space Shuttle, add flexibility to the space program, and free the Shuttle for Shuttle-unique scientific and national security missions. In addition, it called for the development of a totally new rocket booster that could replace the Shuttle about the turn of the century. This launcher would, Quayle noted, fulfill the promise of "routine access to space" that had been intended with the Shuttle. Debate has been intense on this issue, for NASA has been hesitant to pursue the costly development of a new booster while still operating the Shuttle as one of the principal missions of the agency.

CHAPTER 10

A TRAJECTORY FOR
THE SPACE PROGRAM

Beginning with the administration of Ronald Reagan in the 1980s the U.S. civil space program entered a period of growth that lasted into the early 1990s. Not only did the Shuttle begin operation, but Reagan and his successor in office, George H. W. Bush (1924–   ) initiated aggressive efforts to develop a space station, explore the planets, and develop applications satellites. Both Reagan and Bush had a propensity for announcing, à la Kennedy's Apollo decision, dramatic space exploration projects. The similarities between these presidential initiatives, however, began and ended with the public declarations. While broad national support was present in 1961 for the Apollo decision—support that was continually nurtured by senior officials at NASA and in other sectors of the government—a consensus in favor of the Reagan and Bush policies was tenuous at best and could not be properly maintained during the years that followed. All of this raised important questions about the future of the U.S. space program in the twenty-first century. In seeking to analyze this future, this chapter will explore several recent projects undertaken by NASA, some of them either troubled or stillborn, and conclude with comments on the direction the U.S. space program appears to be going.

THE SPACE STATION PROGRAM. In 1984, as part of its interest in reinvigorating the space program, the Reagan administration called for the development of a permanently occupied space station. In a "Kennedyesque" moment, Reagan declared that "America has always been greatest when we dared to be great. We can reach for greatness again. We can follow our dreams to distant stars, living and working in space for peaceful, economic, and scientific gain. Tonight I am directing NASA to develop a permanently manned space station and to do it within a decade." (*See Reading No. 23.*) Congress made a down payment of $150 million for Space Station Freedom in the fiscal year 1985 NASA budget.

The dream of a permanently occupied space station had been omnipresent within NASA since the 1950s, when it had been envisioned as a necessary outpost in the new frontier of space. Most important, it was a jumping off point to the Moon and the planets. Studies of space station configurations had been an important part of NASA planning in the 1960s, since a station met the needs of the agency for an orbital laboratory, observatory, industrial plant, launching platform, and drydock. The station, however, had been forced to the bottom of the priority heap in the early 1970s as NASA went on to build the Space Shuttle, a vehicle that had originally been intended merely as a logistical craft to travel between Earth and the space station.

When the Shuttle first flew in April 1981, the space station reemerged as *the* priority program in human spaceflight. Within three years space policymakers had persuaded Reagan of its importance, and NASA began work on it in earnest. In 1985 the space agency came forward with an $8 billion dual-keel space station configuration, to which were attached a large solar power plant and several modules for microgravity experimentation, life science, technical activities, and habitation. This station also had the capacity for significant expansion through the addition of other modules.

From the outset, both the Reagan administration and NASA intended Space Station Freedom to be an international program. Although a range of international cooperative activities had been carried out in the past—Spacelab, the Apollo-Soyuz Test Project, and scientific data exchange—the station offered an opportunity for a truly integrated effort. The inclusion of international partners, many now with their own rapidly developing spaceflight capabilities, could enhance the effort. In addition, every partnership brought greater legitimacy to the overall program and might help to insulate it from drastic budgetary and political changes. Inciting an international incident because of a change to the station was something neither U.S. diplomats nor politicians relished, and that fact, it was thought, could help stabilize funding, schedule, or other factors that might otherwise be changed in response to short-term political needs.

NASA leaders understood these positive factors, but recognized that international partners would also dilute their authority to execute the program as they saw fit. Assigning an essentially equal partner responsibility for the development of a critical subsystem meant giving up the right to make changes, to dictate solutions, and to control schedules and other factors. In addition to this concern, some expressed fear that bringing foreign partners into the project really meant giving rival nations technical knowledge that only the United States held. No other nation had built a space station on a par with Freedom, and only a handful had a genuine launch capability. Many government officials questioned the advisability of reducing America's technological lead. The control of technology transfer in the international arena was an especially important issue to be considered.

In spite of these concerns, NASA leaders pressed forward with international agreements among 13 nations to take part in the Space Station Freedom program. Japan, Canada, and the nations pooling their resources in the European Space Agency (ESA) agreed in the spring of 1985 to participate. Canada, for instance, decided to build a remote servicing system. Building on its Spacelab experience, ESA agreed to build an attached pressurized science module and an astronaut-tended free-flyer. Japan's contribution was the development and commercial use of an experiment module for materials processing, life sciences, and technologal development. These separate components, with their "plug-in" capacity, eased somewhat the overall management (and congressional) concern about unwanted technology transfer.

Almost from the outset, the Space Station Freedom program was controversial. Most of the debate centered on its costs versus its benefits. One NASA official remembered that "I reached the scream level at about $9 billion," referring to how much U.S. politicians appeared willing to spend on the station. As a result, NASA designed the project to fit an $8 billion research and development funding profile. For many reasons, some of them associated with tough Washington politics, within five years the projected costs had more than tripled and the station had become too expensive to fund fully in an environment in which the national debt had exploded in the 1980s.

NASA pared away at the station budget, in the process eliminating functions that some of its constituencies wanted. This led to a rebellion among some former supporters. For instance, the space science community began complaining that the space station configuration under development did not provide sufficient experimental opportunity. Thomas M. Donahue (1921–   ), an atmospheric scientist from the University of Michigan and chair of the National Academy of Sciences' Space Science Board, commented in the mid-1980s that his group "sees no scientific need for this space station during the next twenty years." He also suggested that "if the decision to build a space station is political and social, we have no problem with that" alluding to the thousands of jobs associated with it. "But don't call it a scientific program."

These concerns were expressed well in the president's 1990 blue-ribbon analysis of the future of the U.S. space program directed by Norman R. Augustine (1935–   ), chief executive officer for Martin Marietta. The report offered several recommendations that cut across the entire spectrum of NASA's organization and activities, but it had some biting commentary on the space station effort. Augustine's commission concluded that "Space Station Freedom be revamped to emphasize life-sciences and human space operations, and include microgravity research as appropriate. It should be reconfigured to reduce cost and complexity." (*See Reading No. 25.*)

Redesigns of Space Station Freedom followed between 1990 and 1993. Each time the project got smaller, less capable of accomplishing the broad projects envisioned for it, less costly, and more controversial. As costs were reduced, capabilities also had to diminish, and increasingly political leaders who had once supported the program questioned its viability. It was a seemingly endless circle, and political wits wondered when the dog would wise up and stop chasing its tail. Some leaders suggested that the nation, NASA, and the overall space exploration effort would be better off if the space station program were terminated. Then, after a few years had passed and additional study and planning had been completed, NASA could come forward with a more viable effort. That Congress did not terminate the program was in part because of the desperate economic situation in the aerospace industry—a result of an overall recession and of military demobilization after the collapse of the Soviet Union and the end of the cold war—and the fact that by 1992 the project had spawned an estimated 75,000 jobs in 39 states, most in California, Alabama, Texas, and Maryland. Politicians were hesitant to kill the station outright because of these jobs, but neither were they willing to fund it at the level required to make it a truly viable program. Barbara Mikulski (1936–   ) (D-MD), chair of the Senate Appropriations subcommittee that handled NASA's budget, summarized this position, "I truly believe that in space station Freedom we are going to generate jobs today and jobs tomorrow—jobs today in terms of the actual manufacturing of space station Freedom, but jobs tomorrow because of what we will learn."

In the latter 1980s and early 1990s a parade of space station managers and NASA administrators, each of them honest in their attempts to rescue the program, wrestled with Freedom and lost. They faced on one side politicians who demanded that the jobs aspect of the project—itself a cause of the overall cost growth—be maintained, and with station users on the other demanding that Freedom's capabilities be maintained, and with people on all sides demanding that costs be reduced. The incompatibility of these various demands ensured that station program management was a task not without difficulties. The NASA administrator since 1 April 1992, Daniel S. Goldin (1940–   ), was faced with a uniquely frustrating situation when these competing claims were made official by the new president, William J. Clinton (1946–   ), who told him in the spring of 1993 to restructure the space station program by reducing its budget, maximizing its scientific use, and ensuring that aerospace industry jobs were not lost. After months of work, NASA came forward with three redesign options for the space station and on 17 June 1993, President Clinton decided to proceed with a moderately priced, moderately capable station design. Near the same time, a dramatically changed international situation allowed NASA to negotiate a landmark decision to include Russia in the building of an international space station. On 7 November 1993 a joint announcement was made by the United States and Russia that they would work together with the other international partners to build a space station for the benefit of all. Even so the space station program remains a difficult issue as the 1990s progress, and public policymakers wrestle with competing political agendas without consensus.

THE SPACE EXPLORATION INITIATIVE. Even more troubling for the space program was another Kennedy-like announcement by President George Bush in 1989 for an ambitious Space Exploration Initiative (SEI) that would return people to the Moon by 2000, establish a lunar base, and, then, using the space station and the Moon, reach Mars by 2010. The price tag for this effort was estimated at a whopping $500 billion over two decades.

In contrast to the space station announcement, this time Congress immediately reacted negatively. In votes for FY 1991 NASA funding, the SEI proposal was virtually zeroed out despite lobbying from Vice President Dan Quayle (1947–   ) as the head of the National Aeronautics and Space Council, an advisory group to the president. Although Bush castigated Congress for not "investing in America's future," members believed such a huge sum could be better spent elsewhere. "We're essentially not doing Moon-Mars," Senator Barbara Mikulski bluntly declared.

In his support of SEI Bush attempted to bring the space program full circle back to the early 1960s, but without the complementary elements that had made Kennedy's Apollo decision viable—that is, a crisis atmosphere that fostered the political will to do something spectacular, a favorable economic and technological climate, and strong public support. In February 1993, looking for ways to cut the federal budget and thereby ease the federal deficit, the new Clinton administration announced that SEI did not fit into its plans for the space program. Accordingly, Goldin said that the agency was not giving up on the idea of sending Americans back to the Moon and on to Mars, but simply "putting it off until we're ready and the nation is able to afford it." The fiscal year 1994 budget contained virtually no funding for SEI, although several smaller robotic programs that would be necessary prior to conducting an ambitious Mars project were still under development.

MISSING COMET HALLEY. In a further example of possible space missions that failed to get off the ground, in

March 1986 the Earth was the closest to Comet Halley that it had been since its last "in-the-vicinity" orbit 76 years earlier. Since this was the first time that technology was present sufficient to send probes to the comet, an armada of five satellites was launched by spacefaring nations to study it. The United States did not sponsor any of these probes, despite its clear leadership in the 1970s in space science, and thus missed an opportunity that would be repeated again only in 2061. Planning for a 1986 mission had begun in 1976, but NASA scientists quickly learned that the cost of the mission would be about $500 million for a basic satellite and much more for more sophisticated probes. Additionally, the development work had to start by 1979 to meet a 1982 launch window for rendezvous. The political will, public desire, and scientific demand for the comet mission was insufficient to secure its political approval. This was complicated by the space science community's bickering over what type of mission to undertake, what instruments to send, and who would direct the effort. The priorities of other space projects in the late 1970s, especially the Shuttle then being developed and several science missions that had *bona fide* scientific merit greater than that proposed for Comet Halley, led the NASA leadership to refrain from pressing for it. A U.S. mission to the comet, therefore, was never undertaken, although one American scientist designed an instrument that was sent on a foreign probe and U.S. space scientists obtained data collected from those other flights.

THE HUBBLE SPACE TELESCOPE. A space science project much in the news in the 1990s, both for positive and negative reasons, was the $2 billion Hubble Space Telescope that had been launched from the Space Shuttle in April 1990. While an earlier project, the Orbiting Astronomical Observatory-1, had been launched in 1968 and placed a telescope above the Earth's obscuring atmosphere, this later project represented a quantum leap forward in astronomical capability. After more than a decade of puritanically funded but productive research and development on the project in the 1970s and early 1980s, NASA began assembling the new space telescope. Through the telescope scientists could gaze farther into space than ever before, viewing galaxies as far away as 15 billion light years. A key component of it was a precision-ground 94-inch primary mirror shaped to within microinches of perfection from ultra-low expansion titanium silicate glass with an aluminum-magnesium fluoride coating.

The Hubble Space Telescope had been scheduled for launch in 1986, but had to be delayed during the Space Shuttle redesign that followed the *Challenger* accident. Excitement abounded as it was finally deployed four years later and the first images began to come back to Earth. The photos provided bright, crisp images against the black background of space, much clearer than pictures of the same target taken by ground-based telescopes. Controllers then began moving the telescope's mirrors to better focus images. Although the focus sharpened slightly, the best image still had a pinpoint of light encircled by a hazy ring or "halo." NASA technicians concluded that the telescope had a "spherical aberration," a mirror defect only 1/25th the width of a human hair, that prevented Hubble from focusing all light to a single point.

At first many believed that the spherical aberration would cripple the 43-foot-long telescope, and NASA received considerable negative publicity, but soon scientists found a way with computer enhancement to work around the abnormality and engineers planned a Shuttle repair mission to fully correct it with an additional instrument. By 1993 the Hubble Space Telescope was returning impressive scientific data on a routine basis. For instance, as recently as 1980, astronomers had believed that an astronomical grouping known as R-136 was a single star, but the Hubble showed that it was made up of more than 60 of the youngest and heaviest stars ever viewed. The dense cluster, located within the Large Magellanic Cloud, was about 160,000 light years from Earth, roughly 5.9 trillion miles away.

Because of the difficulties with the mirror of the Hubble Space Telescope, in December 1993 NASA launched the shuttle *Endeavour* on a repair mission to insert corrective lenses into the telescope and to service other instruments. During a week-long mission, *Endeavour's* astronauts conducted a record five spacewalks and successfully completed all programmed repairs to the spacecraft. The first reports from the Hubble spacecraft indicated that the images being returned were afterward more than an order of magnitude greater than those obtained before.

PLANETARY PROBES. In the late 1980s a new generation of planetary exploration began. Numerous projects came to fruition during the period. For example, the highly successful Magellan mission to Venus by 1993 had provided significant scientific data about the planet. Another such project was the troubled Galileo mission to Jupiter, which even before reaching its destination had become a source of great concern by both NASA and public officials becuase not all of its systems were working properly and many people feared that it would be unable to accomplish its mission. Finally, the Mars Observer was launched and reached its destination in 1993.

The Magellan spacecraft set out for Venus in 1989 to map the surface from orbit with imaging radar. This mission followed the Pioneer 12 spacecraft that had been orbiting the planet for more than a decade, completing a low-resolution radar topographic map, and Pioneer 13, which had dispatched heat-resisting probes to penetrate the atmosphere and communicate information about the surface, the dense clouds, and the 900 degree Fahrenheit temperature. It also built on the work of the Soviet Union, which had compiled radar images of the northern part of Venus and had deployed balloons into the Venusian atmosphere. Magellan arrived at Venus in September 1990 and mapped 99 percent of the surface at high resolution, parts of it in stereo. The amount of digital imaging data the spacecraft returned was more than twice the sum of all returns from previous missions. This data provided some surprises: among them the discovery that plate tectonics was at work on Venus and that lava flows showed clearly the evidence of volcanic activity. In 1993, at the end of its mission, NASA's Jet Propulsion Laboratory shut down the major functions of the Magellan spacecraft, and scientists turned their attention to a detailed analysis of its data.

In October 1989, NASA's Galileo spacecraft began a gravity-assisted journey to Jupiter, where it would send a probe into the atmosphere and observe the planet and its satellites for two years beginning in 1995. Jupiter was of great interest to scientists because it appeared to contain material in its original state left over from the formation of the solar system, and the mission was designed to investigate the chemical composition and physical state of Jupiter's atmosphere and satellites. On the way to Jupiter Galileo encountered both Venus and the Earth and made the first close flyby of asteroid Gaspra in 1991, providing scientific data on all. But the mission was star-crossed. Soon after deployment from the Space Shuttle, NASA engineers learned that Galileo's umbrella-like, high-gain antenna could not be fully deployed. Without this antenna, communication with the spacecraft was both more difficult and time-consuming, and data transmission was greatly hampered. The engineering team working on the project tried a series of cooling exercises designed to shrink the antenna central tower and enable its deployment. Over a period of several months they worked on this maneuver repeatedly, but were unable to free the antenna. Through the end of June 1993 the spacecraft's performance and condition were excellent except that the high-gain antenna was still only partly deployed; science and engineering data were therefore being transmitted via the much slower and less effective low-gain antenna, which the mission team had decided to use for the Jupiter encounter.

The Mars Observer was launched at 1:05 p.m. on Friday, 25 September 1992, from Cape Canaveral, Florida, for an epic-making flight to the Red Planet. The spacecraft was to provide the most detailed data available about Mars as it orbited the planet since what had been collected by the Viking probes of the mid-1970s. The mission was progressing smoothly until about 9 p.m. on Saturday, 21 August 1993, three days before the spacecraft's capture in orbit around Mars, when controllers lost contact with it. The engineering team working on the project at the Jet Propulsion Laboratory responded with a series of commands to turn on the spacecraft's transmitter and to point the spacecraft's antennas toward Earth. No signal came from the spacecraft, however, and the Mars Observer was not heard from again. The loss of the nearly $1 billion Mars Observer probably came as a result of an explosion in the propulsion system's tanks as they were pressurized. This accident caused a significant amount of negative publicity for the space agency, and raised questions about managerial and engineering excellence in NASA. With no response from the Mars Observer, on 29

October 1993, flight controllers concluded scheduled operations.

The problems of the Hubble Space Telescope, Galileo, and Mars Observer also brought to a head the question of scale that had been a perennial source of discussion in satellite programs. Should NASA build a large number and variety of small, inexpensive satellites or consolidate many kinds of experiments onto a few large, expensive spacecraft? Both sides had valid rationales. Small, inexpensive satellites could not accomplish a great deal at any one time and had limited scientific value. Large, costly satellites were a scientist's (but not an accountant's) dream provided they worked properly, but if any component failed, the returns could be greatly diminished. In the 1990s the NASA administrator began to urge a new philosophy of "smaller, cheaper, faster" for the agency's space probes, and advocated a mixture of large and small spacecraft to avoid the long hiatus that came if a mission failed. Such an approach, while also having drawbacks, was designed to minimize the potential loss from such difficulties as those encountered by the space telescope, Mars Observer, and Galileo.

SATELLITE APPLICATIONS. Since the 1960s NASA has been heavily involved in developing and operating various types of applications satellites for communication, weather, navigation, or Earth resources data collection. Communication satellites, some of which operated in geosynchronous orbit, clearly revolutionized the way in which humans reached each other. Since the first Telstar, launched in 1962, it has become possible to send radio messages, telephone calls, and later television programs instantly between distant parts of Earth. Closely akin to communications satellites were navigation satellites that were first placed in orbit in the 1960s to help pilots and sailors find their exact positions in all kinds of weather. Navigators could use these stationary satellites to find their position much as they had used stars in the past, but instead of taking sightings they could listen to satellite radio signals. In the late 1980s a major breakthrough in this technology came with the launching of the Global Positioning System (GPS), a space-based, radio-navigation system for highly precise, worldwide, three-dimensional position, velocity, and timing of vehicles. Through 1992, approximately 100,000 terminals for GPS had been placed in operation, and collectively the system could position vehicles within 16 meters for military users and 100 meters for civilian users. This capability was expected to lead to a significant reorientation in the way navigation would be carried out in the next century.

Meteorological satellites also helped scientists to forecast and study weather and climate. The first of these satellites, the polar-orbiting Tiros, had been launched in the 1960s, but in the thirty years since first employed, meteorological satellites have fundamentally altered both scientific exploration and practical application. With the success of Tiros, NASA and the National Oceanographic and Atmospheric Administration (NOAA) created a second-generation research satellite called Nimbus. More complex than Tiros, Nimbus satellites carried advanced TV cloud-mapping cameras and an infrared radiometer that allowed pictures at night for the first time. Seven Nimbus satellites were placed in orbit between 1964 and 1978, creating the capability to observe the planet 24 hours per day.

When Mt. St. Helens erupted on 18 May 1980, for example, weather satellites tracked the tons of volcanic ash that spread eastwardly, allowing meteorologists both to warn of danger and to study the effects of the explosion on the world's climate. More spectacular, and ultimately more disconcerting, Nimbus 7, in orbit since 1978, revealed that ozone levels over the Antarctic had been dropping for years and had reached record lows by October 1991. This data, combined with that from other sources, led to the 1992 decision to enact U.S. legislation banning chemicals that depleted the ozone layer. In the 1980s NASA and NOAA also began developing the Geostationary Operational Environmental Satellite (GOES) system, which viewed the entire Earth every 30 minutes, day and night, and placed seven GOES into orbit. As the 1990s began, a series of five new satellites, designated GOES-I through -M, was under development by NASA and NOAA for use beyond the year 2000.

In addition to communications, navigation, and weather satellites, NASA got heavily involved in the 1970s in Earth resource mapping satellites, the first of which was the LANDSAT series.

More recently, NASA's Earth Observing System (EOS)—to be launched beginning in 1998—has been under development to explore how the Earth's atmosphere, land, ocean, and life interact by making a variety of simultaneous measurements.

ASSESSMENT. As the U.S. space program entered the last decade of the twentieth century, its reputation had been tarnished by politically-charged debates such as those swirling around the space station and SEI, by the lack of a Comet Halley mission, the mispercieved failure of the Hubble Space Telescope, and the possible deficiency of Galileo, to say nothing of the aftermath of the Shuttle accident. The fate of the space station program was undecided in 1993, but both it and the stillborn Space Exploration Initiative and Comet Halley projects, point up the difficulty of building the constituency for large science and technology programs in a democracy. The rocky course of these projects provide important lessons about the nature of high-technology public policy in modern America. They were striking examples of what social scientists have called heterogeneous engineering, a concept that recognizes that scientific and technological issues are simultaneously organizational, social, economic, and political. Various interests often clash in the decision-making process as difficult calculations have to be made. What perhaps should be suggested is that a complex web or system of ties existed between various people, institutions, and interests all pushing for or opposing the various projects as they emerged in the early 1990s. These interests could come together to make it possible to develop a project that would satisfy the majority of the priorities brought into the political process, but if so many others would undoubtedly be left unsatisfied. In the case of SEI and Comet Halley, no strong coalition of interests ever emerged and the result was a quick closure to the effort.

Even so, NASA could still point to several scientifically important and popularly exciting efforts in space. Many of these revolved around the furtherance of knowledge about space and the universe and the exploitation of space-based technology for practical applications. Its space science program was still the strongest in the world and several projects were underway to continue that leadership. Its satellite application programs were also providing weather, communication, Earth resources, and climatological data of all types for the use both of scientists and public policymakers concerned with crop forecasting, ozone depletion, and a host of other issues. These commercially viable and well-supported programs undoubtedly would survive regardless of any possible retreat from other aspects of space flight.

The president's national space policy reaffirmed in the 1980s and 1990s that efforts outside the Earth's atmosphere served a variety of vital national goals and objectives. Accordingly, human spaceflight, satellite probes, and space-based applications technology all continued to play important roles. At the same time, the nation's leaders had to reevaluate the posture and the long-term direction of the U.S. civilian space program. They found that the civil space exploration effort was essentially discretionary, and that NASA had to maintain a balanced space program within strict fiscal limits. As the second millennium of the common era approached, therefore, NASA leaders started to make hard choices about what would be done how and on what time schedule. The high-priced projects, especially Space Station Freedom and SEI, were seemingly constantly scrutinized and pared back until they bore little resemblance to what had been initially proposed. The space science missions were also reviewed, but there was considerable support for their continuation, provided they were not too costly and had firm backing from the scientific community. Everyone, however, pointed to the benefits derived from such applications technology as weather, navigation, communications, and Earth resources satellites. NASA budgets, which stabilized in 1993 instead of rising as they had in the recent past, reflected this difficult fiscal environment, and while Shuttle and space station declined in public approbation, space science and applications continued to grow.

PART II

READINGS

READING NO. 1

EARLY U.S. PLANS FOR
AN ORBITAL SPACECRAFT[1]

*Just after the end of World War II the newly formed RAND study
group was asked by the U.S. Air Force to investigate the
possibility of launching an orbital satellite, the first serious
study of the subject in the United States. The resulting report,
released in May 1946, suggested that satellites had broad uses
in meteorology, reconnaissance, and communications. But
while extensive in scope and providing much new information on
the value of satellites, including the possibility of a vehicle
which could carry humans, the report was not acted upon. These
excerpts give a sense of the broader thinking that guided the
engineering analyses that comprised the bulk of the document,
although neither the cost of $150 million nor the time frame for
construction of an orbital vehicle proved correct.*

γ   γ   γ

In this report, we have undertaken a conservative and realistic
engineering appraisal of the possibilities of building a spaceship
which will circle the earth as a satellite. The work has been
based on our present state of technological advancement and has
not included such possible future developments as atomic energy.

If a vehicle can be accelerated to a speed of about 17,000
m.p.h. and aimed properly, it will revolve on a great circle path
above the earth's atmosphere as a new satellite. The centrifugal
force will just balance the pull of gravity. Such a vehicle will
make a complete circuit of the earth in approximately 1½ hours.
Of all the possible orbits, most of them will not pass over the
same ground stations on successive circuits because the earth

---

[1] "Preliminary Design of an Experimental World-Circling Spaceship," Douglas
Aircraft Company, Inc., Report No. SM-11827, 2 May 1946, pp. 1–16, Archives, The RAND Corporation, Santa Monica, CA.

will turn about ¹⁄₁₆ of a turn under the orbit during each circuit. The equator is the only such repeating path and consequently is recommended for early attempts at establishing satellites so that a single set of telemetering stations may be used.

Such a vehicle will undoubtedly prove to be of great military value. However, the present study was centered around a vehicle to be used in obtaining much desired scientific information on cosmic rays, gravitation, geophysics, terrestrial magnetism, astronomy, meteorology, and properties of the upper atmosphere. For this purpose, a payload of 500 lbs. and 20 cu ft. was selected as a reasonable estimate of the requirements for scientific apparatus capable of obtaining results sufficiently far-reaching to make the undertaking worthwhile. It was found necessary to establish the orbit at an altitude of about 300 miles to insure sufficiently low drag so that the vehicle could travel for 10 days or more, without power, before losing satellite speed.

The only type of power plant capable of accelerating a vehicle to a speed of 17,000 m.p.h. on the outer limits of the atmosphere is the rocket. The two most important performance characteristics of a rocket vehicle are the exhaust velocity of the rocket and the ratio of the weight of propellants to the gross weight. Very careful studies were made to establish engineering estimates of the values that can be obtained for these two characteristics.

The study of rocket performance indicated that while liquid hydrogen ranks highest among fuels having large exhaust velocities, its low density, low temperature and wide explosive range cause great trouble in engineering design. On the other hand, alcohol though having a lower exhaust velocity, has the benefit of extensive development in the German V-2. Consequently it was decided to conduct parallel preliminary design studies of vehicles using liquid hydrogen-liquid oxygen and alcohol-liquid oxygen as propellants. . . .

Using the above results, it was found that neither hydrogen-oxygen nor alcohol-oxygen is capable of accelerating a single unassisted vehicle to orbital speeds. By the use of a multi-stage rocket, these velocities can be attained by vehicles feasible within the limits of our present knowledge. To illustrate the concept of a multi-stage rocket, first consider a vehicle composed of two parts. The primary vehicle, complete with its rocket motor, tanks, propellants and controls is carried along as the "payload" of a similar vehicle of much greater size. The rocket of the large vehicle is used to accelerate the combination to as great a speed as possible, after which, the large vehicle is discarded and the small vehicle accelerates under its own power, adding its velocity increase to that of the large vehicle. By this means we have obtained an effective decrease in the amount of structural weight that must be accelerated to high speeds. This same idea can be used in designing vehicles with a greater number of stages. A careful analysis of the advantages of staging showed that for a given set of performance require-ments, an optimum number of stages exists. If the stages are too few in number, the required velocities can be attained only by the undesirable process of exchanging payload for fuel. If they are too many, the multiplication of tanks, motors, etc. elimi-nates any possible gain in the effective weight ratio. For the alcohol-oxygen rocket it was found that four stages were best. For the hydrogen-oxygen rocket, preliminary analysis indicated that the best choice for the number of stages was two, but refinements showed the optimum number of stages was three. Unfortunately, insufficient time was available to change the design, so the work on the hydrogen-oxygen was completed using two stages. . . .

It was found that the vehicle could best be guided during its accelerated flight by mounting control surfaces in the rocket jets and rotating the entire vehicle so that lateral components of the jet thrust could be used to produce the desired control forces. It is planned to fire the rocket vertically upward for several miles and then gradually curve the flight path over in the direction in which it is desired that the vehicle shall travel. In order to establish the vehicle on an orbit at an altitude of about 300 miles without using excessive amounts of control it was found desir-able to allow the vehicle to coast without thrust on an extended elliptic arc just preceding the firing of the rocket of the last stage. As the vehicle approaches the summit of this arc, which is at the final altitude, the rocket of the last stage is fired and the vehicle is accelerated so that it becomes a freely revolving satellite.

It was shown that excessive amounts of rocket propellants are required to make corrections if the orbit is incorrectly estab-lished in direction or in velocity. Therefore, considerable atten-tion was devoted to the stability and control problem during the acceleration to orbital speeds. It was concluded that the orbit could be established with sufficient precision so that the vehicle would not inadvertently re-enter the atmosphere because of an eccentric orbit.

Once the vehicle has been established on its orbit, the questions arise as to what are the possibilities of damage by meteorites, what temperatures will it experience, and can its orientation in space be controlled? Although the probability of being hit by very small meteorites is great, it was found that by using reasonable thickness plating, adequate protection could be obtained against all meteorites up to a size where the frequency of occurrence was very small. The temperatures of the satellite vehicle will range from about 40°F when it is on the side of the earth facing the sun to about -20°F when it is in the earth's shadow. Either small flywheels or small jets of compressed gas appear to offer feasible methods of controlling the vehicle's orientation after the cessation of rocket thrust.

An investigation was made of the possibility of safely landing the vehicle without allowing it to enter the atmosphere at such great speeds that it would be destroyed by the heat of air resistance. It was found that by the use of wings on the small final vehicle, the rate of descent could be controlled so that the heat would be dissipated by radiation at temperatures the structure could safely withstand. These same wings could be used to land the vehicle on the surface of the earth.

An interesting outcome of the study is that the maximum acceleration and temperatures can be kept within limits which can be safely withstood by a human being. Since the vehicle is not likely to be damaged by meteorites and can be safely brought back to earth, there is good reason to hope that future satellite vehicles will be built to carry human beings.

It has been estimated that to design, construct and launch a satellite vehicle will cost about $150,000,000. Such an undertaking could be accomplished in approximately 5 years time. The launching would probably be made from one of the Pacific islands near the equator. A series of telemetering stations would be established around the equator to obtain the data from the scientific apparatus contained in the vehicle. The first vehicles will probably be allowed to burn up on plunging back into the atmosphere. Later vehicles will be designed so that they can be brought back to earth. Such vehicles can be used either as long range missiles or for carrying human beings.

READING NO. 2

A REPORT ON
THE "SATELLITE PROBLEM"[1]

In 1952, President Harry S. Truman requested Aristid V. Grosse, a physicist at Temple University who had worked on the Manhattan Project, to study the "satellite problem." The report was finished after Truman left office, but Grosse delivered it to representatives of the Eisenhower administration on 24 September 1953. Grosse's report led directly to the initiation of Project Vanguard, the first U.S. orbital satellite development program. The report also represents the first time that the potential propaganda consequences of a Soviet launch of a satellite were reported directly to top leaders in the U.S. government.

γ   γ   γ

The Present Status of the Satellite Problem

A satellite is a man-made or artificial moon which will rotate around the earth beyond the furthermost extent of its atmosphere, for many years or indefinitely. After it has once reached its orbital velocity the centrifugal force of its motion is held in exact balance by the gravitational attraction of the earth; thus the satellite once on its orbit around the earth *does not require* any additional power to keep it there. Usually altitudes of 300 to 1000 miles above the earth's surface are considered.

As an example, at an altitude of 346 miles above sea level the time necessary for the satellite to travel once around the earth, i.e. its period of revolution, will be exactly 96 minutes or 15 revolutions per day. Its orbital velocity will then be 4.71 miles per second. Similarly, a satellite at an altitude of 1037 miles

---

[1] Aristid V. Grosse, "Report on the Present Status of the Satellite Problem," 25 August 1953, The Research Institute of Temple University, Philadelphia, PA, copy available in NASA Historical Reference Collection, NASA Headquarters, Washington, DC.

above sea level will have a period of revolution of exactly 120 minutes or 12 revolutions per day and a velocity of 4.37 miles per second.

The satellite could be made to travel over the surface of the earth in a wavy line so that in the course of a few successive days most of the North American continent, Europe, Africa and Asia, could be observed from it.

It could be made visible to the naked eye, under clear atmospheric conditions, at dawn and at dusk as a bright fast moving star.

The technical problem of creating a satellite should logically be divided into the two following steps:

1. The unmanned satellite and
2. The manned satellite.

The accomplishment of the first step, in the opinion of even the most skeptical engineers, is possible with the present know-how and engineering knowledge. Since it is not manned by human beings it would not require any essentially new research and development.

A satellite of about 30 feet in length would require the stepping up of the German V-2 rocket by a take-off weight factor of 6-7. This would require essentially the addition of a third large stage to the present well known two stage rockets such as the WAC Corporal mounted on a V-2, which reached an altitude of 250 miles at the White Sands Proving Grounds in February 1949. A design for such a large stage was already on the drawing boards of Dr. von Braun and his associates in Peenemunde, Germany, in 1945. This German project "A-9 + A-10" was designed for transatlantic bombing of the United States. The A-9 stage was a slightly enlarged V-2 (take-off weight 16.3 metric tons vs. 12.8 tons of the V-2) whereas the A-10 stage had a take-off weight of 69 metric tons. Such a three stage rocket would use conventional fuels giving a specific thrust of 220–240 seconds (for example, liquid oxygen + ethyl alcohol 75%, water 25% = 239 seconds, red fuming nitric acid + aniline = 221 seconds). Conventional combustion chambers, pumps, tanks, ignition devices, etc., could be used.

Research scientists have recently demonstrated that much larger specific thrusts can be realized. For example, a liquid fluorine-liquid hydrogen rocket motor can generate a thrust of about 380 seconds. This would permit the use of a much smaller rocket to achieve satellite velocity. However, the engineers feel that this advantage is offset by the necessity of doing a lot of additional research and development in order to bring the high thrust rocket motors and their accessories to the same stage of reliability and smoothness of operation as the conventional rockets. All of this new development would thus cause a loss in time. This would be unwise because it is felt by all engineers that the present rocket fuels and motors will be able to do the satellite job.

The second step or a satellite manned by human beings is decidedly a much more difficult problem. Ultimately, if solved, it would mean the beginning of man's conquest of interstellar space and would have infinite possibilities for the human race. The solution of this problem, however, involves overcoming all the obstacles in the way of man's existence in the vacuum of outer space. It means the overcoming of the absence of a gravitational field on the functions of the human body and the effects of cosmic radiation on it. Although all of these problems have a possibility of ultimate solution, it would require at least a 20-fold expense of human effort, money and time, as compared to Step 1, coupled with an inestimable amount of human ingenuity and invention.

It is felt that the accomplishment of the first step would help solve many of the problems of the second. This writer feels that probably after the successful launching of the first unmanned satellite, a number of such unmanned satellites will be in existence at various altitudes above the surface of the earth, for various purposes. It is thus felt that at this time the main effort should be directed toward solving the unmanned satellite problem.

The value of an unmanned satellite would fall into the following categories.

c)   *Scientific*—with proper electronic and telemetering equipment and devices it would enable us to obtain valuable scientific information regarding the various physical conditions existing in outer space. The satellite would need a concentrated source of energy, which should be *light in weight* and should produce *power for a number of years*. It is considered that such a power plant could be produced by using alpha-active radioactive substances of an average life of a few years in concentrated form, if the appropriate resources of the Atomic Energy Commission could be mobilized.

b) *Military*—again, with the equipment referred to above coupled with televising devices, a satellite station could be a valuable observation post.

c) *Psychological*—with appropriate signaling or broadcasting devices such a satellite could develop into a highly effective sky messenger of the free world.

In the opinion of this writer the last item, i.e., the psychological effect, would be considered of utmost value by the members of the Soviet Politbureau. They would recognize that in the case of atomic and hydrogen bombs the people of the belligerent countries would be subjected to their effects only after the die of World War III is already cast. On the other hand, the satellite would have the enormous advantage of influencing the minds of millions of people the world over during the so-called period of "cold war" or during the peace years *preceding* a possible World War III. In the countries of Asia, where the star gazer since time immemorial has been influencing his countrymen, the spectacle of a man-made satellite would make a profound impression on the minds of the people. The Soviet Union has demonstrated that it has been able to develop the atomic bomb and recently to follow that up with the accomplishment of a thermonuclear reaction on August 12, 1953, as confirmed by the Chairman of the U.S. Atomic Energy Commission, Admiral L. Strauss, much faster than had been generally expected by our scientists and engineers. The building of an unmanned satellite would be a feat of much smaller magnitude than the construction of an atomic bomb since all the basic information was available to the Germans at the end of World War II and is since known both to this country and to the Soviet Union. Furthermore, the industrial plant necessary for the construction of a satellite is much simpler and is now being developed for the guided missiles programs in both countries.

In the Soviet Union the construction of a satellite would amount to only a fraction of the cost in this country, a) because of the use of cheap or slave labor; b) no necessity for great safety precautions, and c) no need for tracking the satellite in the early stages of its flight.

Since the Soviet Union has been following us in the atomic and hydrogen bomb developments, it should not be excluded that the Politbureau might like to take the *lead* in the development of a satellite. They may also decide to dispense with a lot of the complicated instrumentation that we would consider necessary to put into our satellite to accomplish the main purpose, namely, of putting a visible satellite into the heavens first. If the Soviet Union should accomplish this ahead of us it would be a serious blow to the technical and engineering prestige of America the world over. It would be used by Soviet propaganda for all it is worth. Of course, the probable reaction of the American people to a Soviet Satellite circling about 300 miles above Washington, New York, Chicago and Los Angeles, would have to be considered.

At the present time our engineering efforts in this field are limited in scope and distributed over various government agencies. It is recommended as a first step in solving the satellite problem that a small but effective committee be set up composed of our top engineers and scientists in the rocket field, with representatives of the Defense and State Departments. This Committee should report to the top levels of our government and should have for its use and evaluation, all data available to our government and industry on this subject. It should report in detail as to what steps should be taken to launch a satellite successfully into outer space and to estimate the cost and time required for such a development. It is felt that if such a committee were in existence and a definite decision taken by our government regarding the construction of a satellite, that it would fire the enthusiasm and imagination of our engineers and scientists and effectively increase our success in the whole field of rockets and guided missiles.

A. V. Grosse

READING NO. 3

THE SOVIET UNION'S ANNOUNCEMENT OF THE ORBITING OF SPUTNIK I[1]

On 4 October 1957 the Soviet Union launched the first earth orbiting satellite to support the scientific research effort undertaken by several nations during the 1957–1958 International Geophysical Year. The Soviets called the satellite "Sputnik" or "fellow traveler" and reported the achievement in a tersely worded press release printed in Pravda. *The United States had also been working on a scientific satellite program, Project Vanguard, but it had not yet launched a satellite.*

γ   γ   γ

For several years scientific research and experimental design work have been conducted in the Soviet Union on the creation of artificial satellites of the earth.

As already reported in the press, the first launching of the satellites in the USSR were planned for realization in accordance with the scientific research program of the International Geophysical Year.

As a result of very intensive work by scientific research institutes and design bureaus the first artificial satellite in the world has been created. On October 4, 1957, this first satellite was successfully launched in the USSR. According to preliminary data, the carrier rocket has imparted to the satellite the required orbital velocity of about 8000 meters per second. At the present time the satellite is describing elliptical trajectories around the earth, and its flight can be observed in the rays of the rising and setting sun with the aid of very simple optical instruments (binoculars, telescopes, etc.).

---

[1] "Announcement of the First Satellite," from *Pravda*, 5 October 1957, in F. J. Krieger, *Behind the Sputniks* (Washington, D.C., Public Affairs Press, 1958), pp. 311–312.

According to calculations which now are being supplemented by direct observations, the satellite will travel at altitudes up to 900 kilometers above the surface of the earth; the time for a complete revolution of the satellite will be one hour and thirty-five minutes; the angle of inclination of its orbit to the equatorial plane is 65 degrees. On October 5 the satellite will pass over the Moscow area twice—at 1:46 a.m. and at 6:42 a.m. Moscow time. Reports about the subsequent movement of the first artificial satellite launched in the USSR on October 4 will be issued regularly by broadcasting stations.

The satellite has a spherical shape 58 centimeters in diameter and weighs 83.6 kilograms. It is equipped with two radio transmitters continuously emitting signals at frequencies of 20.005 and 40.002 megacycles per second (wave lengths of about 15 and 7.5 meters, respectively). The power of the transmitters ensures reliable reception of the signals by a broad range of radio amateurs. The signals have the form of telegraph pulses of about 0.3 second's duration with a pause of the same duration. The signal of one frequency is sent during the pause in the signal of the other frequency.

Scientific stations located at various points in the Soviet Union are tracking the satellite and determining the elements of its trajectory. Since the density of the rarefied upper layers of the atmosphere is not accurately known, there are no data at present for the precise determination of the satellite's lifetime and of the point of its entry into the dense layers of the atmosphere. Calculations have shown that owing to the tremendous velocity of the satellite, at the end of its existence it will burn up on reaching the dense layers of the atmosphere at an altitude of several tens of kilometers.

As early as the end of the nineteenth century the possibility of realizing cosmic flights by means of rockets was first scientifically substantiated in Russia by the works of the outstanding Russian scientist K[onstantin] E. Tsiolkovskii [Tsiolkovsky].

The successful launching of the first man-made earth satellite makes a most important contribution to the treasure-house of world science and culture. The scientific experiment accomplished at such a great height is of tremendous importance for learning the properties of cosmic space and for studying the earth as a planet of our solar system.

During the International Geophysical Year the Soviet Union proposes launching several more artificial earth satellites. These subsequent satellites will be larger and heavier and they will be used to carry out programs of scientific research.

Artificial earth satellites will pave the way to interplanetary travel and, apparently, our contemporaries will witness how the freed and conscientious labor of the people of the new socialist society makes the most daring dreams of mankind a reality.

READING NO. 4

PLANNING A NATIONAL SPACE AGENCY[1]

The Soviets orbited Sputnik I four months prior to the 4 February 1958 meeting discussed here, but the issue was still much on the mind of members of the Eisenhower administration. By this time, however, it was all but certain that a new space agency would be created. Its exact responsibilities, form, and location, however, were still undecided. Many of these questions are debated in the meeting minutes that follow. The question of the military or civilian character of a new agency was also discussed in a regularly scheduled meeting between the president, vice-president, other White House officials, and Republican leaders in the Congress. The issue was raised in response to the impending reorganization of the Department of Defense, necessitated in part by the increasing sophistication and cost of weapons systems. Missiles and other space related hardware were responsible for a significant portion of the technological revolution sweeping the military services at the time. At this time (February 1958), President Eisenhower had apparently not yet decided that most of the U.S. space program should be carried out under civilian auspices. These minutes provide an important glimpse into the U.S. policymaking process that took place in response to Sputnik I.

γ   γ   γ

OUTER SPACE PROGRAM. A question was raised as to whether a new Space Agency should be set up within Department of Defense (as provided in the pending Defense appropriation bill), or be set up as an independent agency. The President's feeling was essentially a desire to avoid duplication, and priority

---

[1]L. Arthur Minnich, Jr., "Legislative Meeting, Supplementary Notes," 4 February 1958, Dwight D. Eisenhower Presidential Papers, Dwight D. Eisenhower Library, Abilene, KS.

for the present would seem to rest with Defense because of paramountcy of defense aspects. However, the President thought that in regard to non-military aspects, Defense could be the operational agent, taking orders from some non-military scientific group. The National Science Foundation, for instance, should not be restricted in any way in its peaceful research.

Dr. Killian had some reservations as to the relative interest and activity of military vs. peaceful aspects, as did the Vice President who thought our posture before the world would be better if non-military research in outer space were carried forward by an agency entirely separate from the military.

There was some discussion of the prospect of a lunar probe. Dr. Killian thought this might be next on the list of Russian efforts. He had some doubt as to whether the United States should at this late date attempt to press a lunar probe, but the question would be fully canvassed by the Science Advisory Committee in the broad survey it had under way. Dr. Killian thought the United States might do a lunar probe in 1960, or perhaps get to it on a crash program by 1959. Sen. Saltonstall had heard, however, that it might even be accomplished in 1958, if pressed hard enough.

Dr. Killian outlined for the leadership the various phases of future development (along the lines of the subsequent press release listing projects in the "soon," "later," and "much later" categories).

Sen. Knowland complained about having to get his information about Space research from the Democratic Senator from Washington (Jackson)—which was just as bad as having to learn from Mr. Symington anything there was to know about the Air Force.

The President was firmly of the opinion that a rule of reason had to be applied to these Space projects—that we couldn't pour unlimited funds into these costly projects where there was nothing of early value to the Nation's security. He recalled the great effort he had made for the Atomic Peace Ship but Congress would not authorize it, even though in his opinion it would have been a very worthwhile project.

And in the present situation, the President mused, he would rather have a good Redstone than be able to hit the moon, for we didn't have any enemies on the moon.

Sen. Knowland pressed the question of hurrying along with a lunar probe, because of the psychological factor. He recalled the great impact of Sputnik, which seemed to negate the impact of our large mutual security program. If we are close enough to doing a probe, he said, we should press it. The President thought it might be OK to go ahead with it if it could be accomplished with some missile already developed or nearly ready, but he didn't want to just rush into an all-out effort on each one of these possible glamor performances without a full appreciation of their great cost. Also, there would have to be a clear determination of what agency would have the responsibility.

The Vice President reverted to the idea of setting up a separate agency for "peaceful" research projects, for the military would be deterred from things that had no military value in sight. The President thought Defense would inevitably be involved since it presently had all the hardware, and he did not want further duplication. He did not preclude having eventually a great Department of Space.

READING NO. 5

PRIORITIES IN SPACE EXPLORATION[1]

In the months after the Sputnik launch, President Eisenhower directed the President's Science Advisory Committee (PSAC) to concentrate its work on the pace and direction of the nation's space program. PSAC focused heavily on the scientific aspects of the space program. With the president's endorsement, on 26 March 1958 it released a report outlining the importance of space activities, but recommended a cautiously measured pace.

γ   γ   γ

What are the principal reasons for undertaking a national space program? What can we expect to gain from space science and exploration? What are the scientific laws and facts and the technological means which it would be helpful to know and understand in reaching sound policy decisions for a United States space program and its management by the Federal Government? This statement seeks to provide brief and introductory answers to these questions.

It is useful to distinguish among four factors which give importance, urgency, and inevitability to the advancement of space technology. The first of these factors is the compelling urge of man to explore and to discover, the thrust of curiosity that leads men to try to go where no one has gone before. Most of the surface of the earth has now been explored and men now turn to the exploration of outer space as their next objective.

Second, there is the defense objective for the development of space technology. We wish to be sure that space is not used

---

[1]President's Science Advisory Committee, Executive Office of the President, "A Statement by the President and the Introduction to Outer Space," 26 March 1958, copy in NASA Historical Reference Collection, NASA Headquarters, Washington, DC.

to endanger our security. If space is to be used for military purposes, we must be prepared to use space to defend ourselves.

Third, there is the factor of national prestige. To be strong and bold in space technology will enhance the prestige of the United States among the peoples of the world and create added confidence in our scientific, technological, industrial, and military strength.

Fourth, space technology affords new opportunities for scientific observation and experiment which will add to our knowledge and understanding of the earth, the solar system, and the universe.

The determination of what our space program should be must take into consideration all four of these objectives. While this statement deals mainly with the use of space for scientific inquiry, we fully recognize the importance of the other three objectives.

In fact it has been the military quest for ultra long-range rockets that has provided man with new machinery so powerful that it can readily put satellites in orbit and, before long, send instruments out to explore the moon and nearby planets. In this way, what was at first a purely military enterprise has opened up an exciting era of exploration that few men, even a decade ago, dreamed would come in this century. . . .

Since the rocket power plants for space exploration are already in existence or being developed for military needs, the cost of additional scientific research, using these rockets, need not be exorbitant. Still, the cost will not be small, either. This raises an important question that scientists and the general public (who will pay the bill) both must face: Since there are still so many unanswered scientific questions and problems all around us on earth, why should we start asking new questions and seeking out new problems in space? How can the results possibly justify the cost?

Scientific research, of course, has never been amenable to rigorous cost accounting in advance. Nor, for that matter, has exploration of any sort. But if we have learned one lesson, it is that research and exploration have a remarkable way of paying off—quite apart from the fact that they demonstrate that man is alive and insatiably curious. And we all feel richer for knowing what explorers and scientists have learned about the universe in which we live.

It is in these terms that we must measure the value of launching satellites and sending rockets into space. . . .

The scientific opportunities are so numerous and so inviting that scientists from many countries will certainly want to participate. Perhaps the International Geophysical Year will suggest a model for the international exploration of space in the years and decades to come.

The timetable . . . suggests the approximate order in which some of the scientific and technical objectives mentioned in this review may be attained.

The timetable is not broken down into years, since there is yet too much uncertainty about the scale of the effort that will be made. The timetable simply lists various types of space investigations and goals under three broad headings: Early, Later, Still Later. . . .

EARLY
1. Physics
2. Geophysics
3. Meteorology
4. Minimal Moon Contact
5. Experimental Communications
6. Space Physiology

LATER
1. Astronomy
2. Extensive Communications
3. Biology
4. Scientific Lunar Investigation
5. Minimal Planetary Contact
6. Human Flight in Orbit

STILL LATER
1. Automated Lunar Exploration
2. Automated Planetary Exploration
3. Human Lunar Exploration and Return

AND MUCH LATER STILL

Human Planetary Exploration

In conclusion, we venture two observations. Research in outer space affords new opportunities in science, but it does not diminish the importance of science on earth. Many of the secrets of the universe will be fathomed in laboratories on earth, and the progress of our science and technology and the welfare of the Nation require that our regular scientific programs go forward without loss of pace, in fact at an increased pace. It would not be in the national interest to exploit space science at the cost of weakening our efforts in other scientific endeavors. This need not happen if we plan our national program for space science and technology as part of a balanced national effort in all science and technology.

Our second observation is prompted by technical considerations. For the present, the rocketry and other equipment used in space technology must usually be employed at the very limit of its capacity. This means that failures of equipment and uncertainties of schedule are to be expected. It therefore appears wise to be cautious and modest in our predictions and pronouncements about future space activities—and quietly bold in our execution.

READING NO. 6

THE CREATION OF NASA[1]

In the aftermath of Sputnik, the Eisenhower administration sponsored legislation creating an American civilian space agency. The White House decided that the new agency would be formed from the National Advisory Committee for Aeronautics (NACA) and rocket and space personnel involved in various defense programs, as needed. On 5 March 1958 Eisenhower approved a final memorandum ordering the Bureau of Budget to draft a space bill. It was ready three weeks later and was sent to Congress on 2 April. Senator Lyndon Johnson had a great deal of influence on the form of the final bill, which was passed after lengthy congressional deliberations and signed into law on 29 July 1958. As a result, the National Aeronautics and Space Administration was established effective 1 October 1958.

γ   γ   γ

An Act

To provide for research into problems of flight within and outside the earth's atmosphere, and for other purposes.

Be it enacted by the Senate and House of Representatives of the United States of America in Congress assembled,

TITLE I—SHORT TITLE, DECLARATION OF POLICY, AND DEFINITIONS

SHORT TITLE

Sec. 101. This act may be cited as the "National Aeronautics and Space Act of 1958".

---

[1] "National Aeronautics and Space Act of 1958," Public Law #85-568, 72 Stat., 426. Signed by the President on 29 July 1958, Record Group 255, National Archives and Records Administration, Washington, DC.

DECLARATION OF POLICY AND PURPOSE

Sec. 102. (a) The Congress hereby declares that it is the policy of the United States that activities in space should be devoted to peaceful purposes for the benefit of all mankind.

(b) The Congress declares that the general welfare and security of the United States require that adequate provision be made for aeronautical and space activities. The Congress further declares that such activities shall be the responsibility of, and shall be directed by, a civilian agency exercising control over aeronautical and space activities sponsored by the United States, except that activities peculiar to or primarily associated with the development of weapons systems, military operations, or the defense of the United States (including the research and development necessary to make effective provision for the defense of the United States) shall be the responsibility of, and shall be directed by, the Department of Defense; and that determination as to which such agency has responsibility for and direction of any such activity shall be made by the President in conformity with section 201 (e).

(c) The aeronautical and space activities of the United States shall be conducted so as to contribute materially to one or more of the following objectives:

(1) The expansion of human knowledge of phenomena in the atmosphere and space;

(2) The improvement of the usefulness, performance, speed, safety, and efficiency of aeronautical and space vehicles;

(3) The development and operation of vehicles capable of carrying instruments, equipment, supplies and living organisms through space;

(4) The establishment of long-range studies of the potential benefits to be gained from, the opportunities for, and the problems involved in the utilization of aeronautical and space activities for peaceful and scientific purposes.

(5) The preservation of the role of the United States as a leader in aeronautical and space science and technology and in the application thereof to the conduct of peaceful activities within and outside the atmosphere.

(6) The making available to agencies directly concerned with national defenses of discoveries that have military value or significance, and the furnishing by such agencies, to the civilian

agency established to direct and control nonmilitary aeronautical and space activities, of information as to discoveries which have value or significance to that agency;

(7) Cooperation by the United States with other nations and groups of nations in work done pursuant to this Act and in the peaceful application of the results, thereof; and

(8) The most effective utilization of the scientific and engineering resources of the United States, with close cooperation among all interested agencies of the United States in order to avoid unnecessary duplication of effort, facilities, and equipment. . . .

TITLE II—COORDINATION OF AERONAUTICAL AND SPACE ACTIVITIES

NATIONAL AERONAUTICS AND SPACE COUNCIL

Sec. 201. (a) There is hereby established the National Aeronautics and Space Council (hereinafter called the "Council") which shall be composed of—

(1) the President (who shall preside over meetings of the Council);

(2) the Secretary of State;

(3) the Secretary of Defense;

(4) the Administrator of the National Aeronautics and Space Administration;

(5) the Chairman of the Atomic Energy Commission;

(6) not more than one additional member appointed by the President from the departments and agencies of the Federal Government; and

(7) not more than three other members appointed by the President, solely on the basis of established records of distinguished achievement from among individuals in private life who are eminent in science, engineering, technology, education, administration, or public affairs.

(b) Each member of the Council from a department or agency of the Federal Government may designate another officer of his department or agency to serve on the Council as his alternate in his unavoidable absence.

(c) Each member of the Council appointed or designated under paragraphs (6) and (7) of subsection (a), and each alternate member designated under subsection (b), shall be appointed or designated to serve as such by and with the advice and consent of the Senate, unless at the time of such appointment or designation he holds an office in the Federal Government to which he was appointed by and with the advice and consent of the Senate.

(d) It shall be the function of the Council to advise the President with respect to the performance of the duties prescribed in subsection (e) of this section.

(e) In conformity with the provisions of section 102 of this Act, it shall be the duty of the President to—

(1) survey all significant aeronautical and space activities, including the policies, plans, programs, and accomplishments of all agencies of the United States engaged in such activities;

(2) develop a comprehensive program of aeronautical and space activities to be conducted by agencies of the United States;

(3) designate and fix responsibility for the direction of major aeronautical and space activities;

(4) provide for effective cooperation between the National Aeronautics and Space Administration and the Department of Defense in all such activities, and specify which of such activities may be carried on concurrently by both such agencies notwithstanding the assignment of primary responsibility therefor to one or the other of such agencies; and

(5) resolve differences arising among departments and agencies of the United States with respect to aeronautical and space activities under this Act, including differences as to whether a particular project is an aeronautical and space activity. . . .

NATIONAL AERONAUTICS AND SPACE ADMINISTRATION

Sec. 202. (a) There is hereby established the National Aeronautics and Space Administration (hereinafter called the "Administration"). The Administration shall be headed by an Administrator, who shall be appointed from civilian life by the President by and with the advice and consent of the Senate, . . . Under the supervision and direction of the President, the Administrator shall be responsible for the exercise of all powers and the discharge of all duties of the Administration, and shall have authority and control over all personnel and activities, thereof. . . .

FUNCTIONS OF THE ADMINISTRATION

Sec. 203. (a) The Administration, in order to carry out the purpose of this Act, shall—

(1) plan, direct, and conduct aeronautical and space activities;

(2) arrange for participation by the scientific community in planning scientific measurements and observations to be made through use of aeronautical and space vehicles, and conduct or arrange for the conduct of such measurements and observations; and

(3) provide for the widest practicable and appropriate dissemination of information concerning its activities and the results thereof. . . .

INTERNATIONAL COOPERATION

Sec. 205. The Administration, under the foreign policy guidance of the President, may engage in a program of international cooperation in work done pursuant to the Act, and in the peaceful application of the results thereof, pursuant to agreements made by the President with the advice and consent of the Senate. . . .

Sec. 301. (a) The National Advisory Committee for Aeronautics, on the effective date of this section, shall cease to exist. On such date all functions, powers, duties, and obligations, and all real and personal property, personnel (other than members of the Committee), funds, and records of that organization, shall be transferred to the Administration. . . .

TRANSFER OF RELATED FUNCTIONS

Sec. 302. (a) Subject to the provisions of this section, the President, for a period of four years after the date of enactment of this Act, may transfer to the Administration any functions (including powers, duties, activities, facilities, and parts of functions) of any other department or agency of the United States, or of any officer or organizational entity thereof, which relate primarily to the functions, powers, and duties of the

Administration as prescribed by section 203 of this Act. In connection with any such transfer, the President may, under this section or other applicable authority, provide for appropriate transfers of records, property, civilian personnel, and funds. . . .

(SAM RAYBURN)
Speaker of the House of Representatives
(RICHARD NIXON)
*Vice President of the United States and
    President of the Senate.*

READING NO. 7

THE BEGINNINGS OF
SPACE SCIENCE POLICY[1]

Immediately after NASA was formed it had to establish a whole series of policy positions to deal with its roles, mission, organization, and operation. One of the most ticklish was its approach toward selecting scientific experiments to fly on NASA rockets, and by extension the scientists favored with NASA support. Several scientists registered complaints about the domineering attitude of NASA project engineers toward experimenters and about the difficulty outside scientists had in competing with NASA scientists. In April 1960, therefore, Homer E. Newell, Assistant Director for Space Sciences, prepared a policy statement establishing a detailed procedure for selecting science activities for NASA missions and provided a description of roles and responsibilities of scientist and project managers. It proved to be a durable procedure. NASA has continued to use the same basic procedure laid down in this document since that time. It represents one of several difficult issues to be decided during the early history of the space agency.

γ   γ   γ

ESTABLISHMENT AND CONDUCT
OF SPACE SCIENCES PROGRAM SELECTION
OF SCIENTIFIC EXPERIMENTS

1. PURPOSE

This Instruction defines responsibilities and establishes procedures for the conduct of the NASA Space Sciences Program.

---

[1]NASA Office of Space Flight Programs, "Establishment and Conduct of Space Sciences Program—Selection of Scientific Experiments," NASA Management Instruction TMI 37-1-1, 15 April 1960; NASA Historical Reference Collection, NASA Headquarters, Washington, DC.

2. BACKGROUND

Under the provisions of the National Aeronautics and Space Act of 1958 (42 U.S.C. 2451 et. seq.), the NASA is responsible for developing and executing a program in space sciences which is scientifically sound and in which the scientific community has broad participation. Success of the program rests in large measure on the ideas and technical abilities of participating scientists, both within and outside NASA. It is essential, therefore, that such competence be utilized in developing and carrying out scientific space missions and experiments, in analyzing research and development requirements, and in recommending efforts to further national program goals.

3. PROGRAM RESPONSIBILITIES

a. *Director of Space Flight Programs.* The Director of Space Flight Programs is responsible for overall direction of the NASA space sciences program, including:

   (1) Establishment of the short and long range scientific program;

   (2) Selection of experiments, experimenters, and specific flight missions;

   (3) Determining research and development needs to meet overall scientific objectives; and

   (4) Appraising results of research efforts.

b. *Space Sciences Steering Committee.* The Space Sciences Steering Committee, appointed by the Director of Space Flight Programs, serves as the focal point for space sciences activities and is responsible for the review and approval for submission to the Director of Space Flight Programs of:

   (1) Proposed short and long range space sciences programs;

   (2) Proposed experiments, experimenters and contractors;

   (3) Program and budgetary breakdowns and supporting research recommendations; and

   (4) Scientific space science assignments for the Goddard Space Flight Center and the Jet Propulsion Laboratory.

c. *Space Sciences Steering Committee Subcommittees.* Subcommittees are appointed by the Director of Space Flight Programs for various space science disciplines or groups of disciplines, including Aeronomy, Ionospheric Physics, Energetic Particles, Astronomy and Solar Physics, Lunar Sciences, and Planetary and Interplanetary Sciences. Such subcommittees serve in an advisory capacity to the Steering Committee and the Assistant Directors of Space Flight Programs and are responsible in their own areas of interest and competence for providing advice and assistance in:

(1) Formulating short and long range plans;
(2) Analyzing, evaluating, and recommending proposed flight experiments and supporting research; and
(3) Reviewing programs for weaknesses, gaps, and imbalances, and recommending necessary actions to correct such inadequacies.

d. *Assistant Directors of Space Flight Programs*. The Assistant Directors of Space Flight Programs are responsible for:

(1) Working with the subcommittees to organize "state-of-the-art" information in pertinent scientific disciplines;
(2) Making tentative selections of experiments and experimenters based on recommendations of subcommittees and field centers;
(3) Working directly with the appropriate field centers to secure the necessary budgetary backup and supporting documentation; and
(4) Supporting and coordinating the research and development work of the field centers in executing approved programs and missions.

e. *Field Centers*. The Goddard Space Flight Center is responsible for conducting missions involving earth satellites and sounding rockets. The Jet Propulsion Laboratory is responsible for conducting unmanned missions involving lunar and deep space probes. In carrying out these responsibilities, such installations will:

(1) Initiate proposals for and participate in the performance of space science experiments and projects;
(2) Technically evaluate proposals submitted by Headquarters for recommendations;
(3) Analyze supporting requirements and recommend scheduling of space sciences programs;

    (4) Prepare and operate, either in-house or by contracts consistent with established policy, the necessary spacecraft to carry out approved scientific missions;

    (5) Conduct or contract for supporting research on advanced technology and instrumentation; and

    (6) Monitoring of selected Headquarters research and development contracts.

4. SELECTION PROCEDURES

    a. Proposed experiments submitted by scientists within and without NASA will be forwarded to the appropriate Assistant Director of Space Flight Programs. The Assistant Directors will submit such proposals to the appropriate advisory subcommittee and to Centers for review and advice. In selecting experiments, proposals from research scientists will be considered on the following basis:

        (1) Desirability within the discipline to which it pertains;

        (2) Probability of acquiring positive scientific results;

        (3) Worth and timeliness in comparison with other competing proposals; and

        (4) Competence and experience of its proposer.

    b. With advice and assistance of the Center and appropriate subcommittees, the Assistant Directors of Space Flight Programs will make tentative selections of experiments and experimenters, and will submit such recommendations to the Space Sciences Steering Committee.

    c. The Space Sciences Steering Committee will review the detailed plans and forward its recommendation, including the designation of the Center to be assigned the technical management responsibility, to the Director of Space Flight Programs for approval.

    d. The Director of Space Flight Programs will approve the mission and will assign the responsibility for program execution.

5. RELATIONSHIP BETWEEN CENTERS AND EXPERIMENTING SCIENTISTS

    a. After selection of flight experiments, funding for prototype models or design concepts of scientific instruments for the selected experiments will be provided either by NASA Headquarters or by the Center with the approval of the Office of Space Flight Programs. At this time, a

Center may be assigned the responsibility for technical monitoring of selected Headquarters contracts. The schedule for completion of prototype models will be established by the Center, consistent with the spacecraft development program.

b. Completed prototypes or design concepts will be delivered to the Center and evaluated by the Center personnel in collaboration with the experimenters. Additional development of the selected instruments will be made under the technical direction of the Center in collaboration with the experimenting scientists. If the Center personnel determine, during the course of fabrication of the flight instruments, that modification of the functional specifications are required in order that the instruments operate reliably in the overall system, such modifications will be made on the basis of agreement between the Center and the experimenters. For modifications which imply major changes in the scientific objectives of the experiment, concurrence of the Office of Space Flight Programs will be obtained by the Center.

c. Based on the functional specifications determined by the responsible experimenters, the following functions will be performed by or under the direction of the Center with the assistance of the experimenters:

   (1) Fabrication;
   (2) Testing;
   (3) Calibration;
   (4) Checkout of flight instruments;
   (5) Integration of experiments into payload and/or spacecraft;
   (6) Participation, as necessary, in field operations; and
   (7) Acquisition and reduction of data from measurements taken in flight.

d. Each selected experimenting scientist will be responsible for:

   (1) Preparing the prototype instruments and associated equipment for his experiments;
   (2) Cooperating in the preparation of flight instrumentation, its environmental testing and calibration for flight;

      (3) Participating, as necessary, in field operations; and

      (4) Analyzing and reporting the data from his experiment.

6. PAYLOAD DESIGN AND FABRICATION

    The responsible Center, with the concurrence of the Office of Space Flight Programs, will determine whether the Center or an outside contractor will design and construct the scientific instrument payload and/or spacecraft.

7. RELATIONSHIPS WITH UNIVERSITIES AND NON-PROFIT ORGANIZATIONS

    a. The Office of Space Flight Programs and Centers will inform each other concerning their plans involving universities and nonprofit organizations and of all concepts and dealings with the scientific community.

    b. With prior approval by the Office of Space Flight Programs, Centers may invite proposals or experiments, including the supplying of flight hardware, from universities and other nonprofit organizations in accordance with overall NASA program objectives. Centers are not authorized to proceed with negotiations for research and development effort with universities and nonprofit organizations without prior approval of the Office of Space Flight Programs.

    c. Proposals received by Centers from universities and nonprofit organizations will be forwarded to the Office of Space Flight Programs for preliminary appraisal and, where approriate, for assignment of detailed technical evaluation. The Director of Space Flight Programs will make the determination whether or not to proceed.

    d. For those proposals which the Office of Space Flight Programs supports, copies of such proposals will be forwarded to the Director, Division of Research Grants and Contracts, Office of Business Administration, NASA Headquarters. The Division of Research Grants and Contracts, NASA Headquarters, will:

      (1) Determine the form of the contractual arrangement to be used, that is contract or grant;

      (2) Make the preliminary contact with the business management of the university or nonprofit organization leading to a contractual arrangement; and

(3) When requested, proceed to negotiate and consummate the contract.

Where the contract is a field assignment, the Office of Space Flight Programs, after obtaining the above clearances, will inform the Center that it is authorized to negotiate and consummate the contract.

READING NO. 8

THE "MAN-IN-SPACE" PROGRAM[1]

When NASA submitted its 1962 budget request to the Bureau of the Budget in May 1960, President Eisenhower learned officially for the first time of the agency's plans for a lunar landing program. He asked his science advisor, George Kistia-kowsky, to study "the goals, the missions and the costs" of the human spaceflight program that NASA had in mind. Kistiakow-sky appointed a six-person study team chaired by Brown University chemistry professor Donald Hornig, and it presented its findings to the president at a 20 December 1960 meeting. Eisenhower has been quoted as saying that he was not willing to "hock his jewels" (referring to the decision by Spanish monarchs Ferdinand and Isabella to finance the initial expedition of Christopher Columbus) to send people to the Moon. The report reflected that basic ideology and viewed a lunar landing program only as a long-term goal.

γ   γ   γ

We have bean plunged into a race for the conquest of outer space. As a reason for this undertaking some look to the new and exciting scientific discoveries which are certain to be made. Others feel the challenge to transport man beyond frontiers he scarcely dared dream about until now. But at present the most impelling reason for our effort has been the international political situation which demands that we demonstrate our technological capabilities if we are to maintain our position of leadership. For all of these reasons we have embarked on a complex and costly adventure. It is the purpose of this report to clarify the goals, the missions and the costs of this effort in the foreseeable

[1]President's Science Advisory Committee, "Report of the Ad Hoc Panel on Man-in-Space," 16 December 1960, Dwight D. Eisenhower Presidential Papers, Dwight D. Eisenhower Library, Abilene, KS.

future, particularly with regard to the man-in-space program. . . .

The initial American attempt to launch a manned capsule into orbital flight, Project Mercury, is already well advanced. It is a somewhat marginal effort, limited by the thrust of the Atlas booster. It has as its goal the launching of a one man capsule into orbit around the earth and its successful return to earth. The fact that the thrust of any available American booster is barely sufficient for the purpose means that it is difficult to achieve a high probability of a successful flight while also providing adequate safety for the Astronaut. Achieving reliability on both accounts will strain our capabilities. A difficult decision will soon be necessary as to when or whether a manned flight should be launched. The chief justification for pushing Project Mercury on the present time scale lies in the political desire either to be the first nation to send a man into orbit, or at least to be a close second.

The marginal capability cannot be changed substantially until the Saturn booster becomes available. The NASA program for utilizing Saturn involves the development of the so called Apollo spacecraft. The Saturn rocket which is being developed now (C-1) should be capable of launching a spacecraft of about 19,000 lbs into a low earth orbit. The proposed Apollo spacecraft weight of 15,000 lbs is well within this limit and would enable orbital qualification flights of the Apollo spacecraft (some manned) about 1966–1968. Such a manned flight would occur after about 25 Saturn C-1s have been tested and much depends on whether a demonstrated reliability can be attained in this rather small number of tests. The Apollo spacecraft, as presently envisioned, would carry three men who would exercise control from within the spacecraft and be able to return to earth within a fairly well defined area. The chief purpose of the early Apollo missions would be to gain experience in manned flight, to learn more of the problems encountered by crews under such new conditions and to aid in the development of a spacecraft for more ambitious missions.

The full capabilities of the Saturn booster cannot be utilized until a large hydrogen-oxygen second stage has been developed. The C-2 Saturn, utilizing the new high-performance stage, is expected to enter the test phase about 1965 and may be available for manned flight in 1968 or 1969. There is again a question as to whether 16 flights will be enough to demonstrate sufficient reliability for its use in manned missions.

The Saturn C-2 is expected to lift about 40,000 lbs into low earth orbit and it is planned to utilize this capability to send up an "orbiting laboratory" capable of staying aloft for two weeks or more. It is our opinion that an orbiting laboratory of this size could produce considerably more scientific information if it were wholly instrumented rather than manned. Alternatively, we believe that the valid scientific missions to be performed by a manned laboratory of this size could be accomplished using a much smaller unmanned instrumented spacecraft which would in turn require a smaller booster system. The large manned orbiting laboratory might be of value as a life sciences laboratory to acquire physiological and psychological data on humans, to study life support mechanisms, to perform biological studies, and to carry out engineering tests under gravity free conditions. In short, its major mission appears to be the preparation for further steps in the manned exploration of space. To take such steps, the Apollo spacecraft may be launched into successively more elliptical orbits which carry it further and further from the earth, culminating about 1970 in a manned flight around the moon and back to the earth. The Apollo program in itself does not reach what might be considered to be the next major goal in manned space flight, i.e. manned landing on the moon. It does, however, appear to represent a logical approach to that goal in that it will develop spacecraft and crews for space flight and will enable us to gain experience in navigation and successful return from increasingly difficult trips. In the meantime it should be possible to obtain far more detailed information about the moon by unmanned spacecraft and lunar landing craft than the crew of the circumlunar flight could gain.

None of the boosters now planned for development are capable of landing on the moon with sufficient auxiliary equipment to return the crew safely to earth. To achieve this goal, a new program much larger than Saturn will be needed. It is likely to take one of three forms:

1. An all-chemical liquid-fueled rocket, the Nova, might be developed to take the trip directly. It would require a booster

with about 6 times the thrust of the Saturn and utilizing either kerosene or hydrogen/oxygen. The upper stage of the Nova would require hydrogen/oxygen and at least one stage would probably be an existing stage from the Saturn development program.

2. If a suitable nuclear upper stage could be developed, the Nova vehicle could conceivably become a combination chemical-nuclear system. This system would still require the development of a first stage chemical booster with thrust of the same order of magnitude as that described for the all chemical system. . . .

3. Rendezvous techniques, utilizing either Saturn C-2 vehicles or some type of advanced Saturn vehicles, could be employed to lift into an earth orbit the hardware and fuel necessary to perform the manned lunar landing mission. In this system, a series of vehicles would be launched into a temporary earth orbit where they would rendezvous to enable fueling of the spacecraft and, if necessary, assembly of the component parts of the spacecraft. This spacecraft would then be used to transport the manned payload to the moon and thence back to Earth. These techniques will require considerable development, and are at present only in a preliminary study phase.

It is clear that any of the routes to land a man on the moon require a development much more ambitious than the present Saturn program. Not only must much bigger boosters probably be developed, but rockets and guidance mechanisms for the safe landing and then for return from moon to earth by means of additional rockets must be developed and tested. Nevertheless, it must be pointed out that this new, major step is implicit in undertaking the proposed manned Saturn program, for the first really big achievement of the man-in-space program would be the lunar landing.

The succeeding step, manned flight to the vicinity of Venus or Mars represents a problem and order of magnitude greater than that involved in the manned lunar landing. Not only does it appear to be insoluble in terms of chemical rockets, thus requiring the development of suitable nuclear rockets or nuclear-powered electric propulsion devices, but it also poses serious problems in terms of life support and radiation shielding for journeys requiring times ranging from many months to years. . . .

Certainly among the major reasons for attending the manned exploration of space are emotional compulsions and national aspirations. These are not subjects which can be discussed on technical grounds. However, it can be asked whether the presence of a man adds to the variety or quality of the observations which can be made from unmanned vehicles, in short whether there is a scientific justification to include man in space vehicles.

It is said that an astronaut's judgment, decision-making capability and resourcefulness can increase the probability of successful accomplishment of a space mission and expand the variety and quality of observations performed. On the other hand, man's senses can be satisfactorily duplicated at remote locations by the use of available instrumentation and advances in the state of the art are continually increasing the ability to transmit information back to a central receiving point. With such an instrumented system, the decisions requiring man's mental capabilities can be performed by many men in a normal environment and with the aid of elaborate computational aids, where necessary.

The following considerations seem pertinent:

1. Information from unmanned flights is a necessary prerequisite to manned flight.

2. The degree of reliability that can be accepted in the entire mechanism is very much less for unmanned than for manned vehicles. As the systems become more complex this may make a decisive difference in what one dares to undertake at any given time.

3. From a purely scientific point of view it should be noted that unmanned flights to a given objective can be undertaken much earlier. Hence repeated observations, changes of objectives and the learning by experience are more feasible.

It seems, therefore, to us at the present time that man-in-space cannot be justified on purely scientific grounds, although more thought may show that there are situations for which this is not true. On the other hand, it may be argued that much of the motivation and drive for the scientific exploration of space is derived from the dream of man's getting into space himself. . . .

Conclusions:

1. The first major goal of the man-in-space program is to orbit a man about the earth. It will cost about 350 million dollars.

2. The next goal, of an intermediate nature, is the manned circum-navigation of the moon. It will cost about 8 billion dollars.

3. The second major goal, landing on the moon, can only be achieved about 1975 after an additional national expenditure in the vicinity of 26 to 38 billion dollars.

4. The Saturn program is a necessary intermediate step toward manned lunar landing but must be followed by a much bigger development before manned lunar landing is possible.

5. The unmanned program is a necessary prerequisite to a manned program. Even if there were no manned program, the unmanned program might yield as much scientific knowledge and on this basis would be justified in its own right.

6. Even if there were no man-in-space program, Saturn C-2 is still a minimum vehicle for close-up instrumented study of Venus and Mars, for unmanned trips to more distant planets, and for putting roving vehicles on the surface of the moon.

7. Manned trips to the vicinity of Venus or Mars are not yet foreseeable.

READING NO. 9

PRESIDENT KENNEDY ASKS THE VICE PRESIDENT FOR RECOMMENDATIONS ON THE U.S. SPACE PROGRAM[1]

The memorandum that follows led directly to the aggressive Apollo program. By posing the question "Is there any . . . space program which promises dramatic results in which we could win?" President Kennedy set in motion a review that concluded that only a crash effort to send Americans to the Moon met the criteria Kennedy had laid out. This memorandum followed a week of discussion within the White House on how best to respond to the challenge to U.S. interests posed by the 12 April 1961 orbital flight of Soviet Cosmonaut Yuri Gagarin.

γ   γ   γ

THE WHITE HOUSE
WASHINGTON

April 20, 1961

MEMORANDUM FOR
    VICE PRESIDENT

In accordance with our conversation I would like for you as Chairman of the Space Council to be in charge of making an overall survey of where we stand in space.

1. Do we have a chance of beating the Soviets by putting a laboratory in space, or by a trip round the moon, or by a rocket to land on the moon, or by a rocket to go to the moon and back with a man. Is there any other space program which promises dramatic results in which we could win?

2. How much additional would it cost?

---

[1] John F. Kennedy, "Memorandum for the Vice President," 20 April 1961, John F. Kennedy Presidential Papers, John F. Kennedy Presidential Library, Boston, MA.

3. Are we working 24 hours a day on existing programs. If not, why not? If not, will you make recommendations to me as to how work can be speeded up.

4. In building large boosters should we put our emphasis on nuclear, chemical or liquid fuel, or a combination of these three?

5. Are we making maximum effort? Are we achieving necessary results?

I have asked Jim Webb, Dr. Wiesner, Secretary McNamara and other responsible officials to cooperate with you fully. I would appreciate a report on this at the earliest possible moment.

[Signed]
John F. Kennedy

READING NO. 10

THE VICE PRESIDENT ANSWERS[1]

This memorandum, signed by Vice President Johnson, was the first report to President Kennedy on the results of the review he had ordered of the space program on 20 April. It identified a human lunar landing by 1966 or 1967 as the first dramatic space project in which the United States could beat the Soviet Union. Vice President Johnson identified U.S. "leadership" in the world arena as sufficient justification of this undertaking in space.

γ   γ   γ

OFFICE OF THE VICE PRESIDENT
WASHINGTON, D.C.

April 28, 1961

MEMORANDUM FOR PRESIDENT
Subject: Evaluation of Space Program.

Reference is to your April 20 memorandum asking certain questions regarding this country's space program.

A detailed survey has not been completed in this time period. The examination will continue. However, what we have obtained so far from knowledgeable and responsible persons makes this summary reply possible.

Among those who have participated in our deliberations have been the Secretary and Deputy Secretary of Defense; General Schriever (AF); Admiral Hayward (Navy); Dr. von Braun (NASA); the Administrator, Deputy Administrator, and other top officials of NASA; the Special Assistant to the President on Science and Technology; representatives of the Director of the Bureau of the Budget; and three outstanding non-Government

[1]Lyndon B. Johnson to the President, "Evaluation of Space Program," 28 April 1961, John F. Kennedy Presidential Files, NASA Historical Reference Collection, NASA Headquarters, Washington, DC.

citizens of the general public: Mr. George Brown (Brown & Root, Houston, Texas); Mr. Donald Cook (American Electric Power Service, New York, N.Y.); and Mr. Frank Stanton (Columbia Broadcasting System, New York, N.Y.).

The following general conclusions can be reported:

a. Largely due to their concentrated efforts and their earlier emphasis upon the development of large rocket engines, the Soviets are ahead of the United States in world prestige attained through impressive technological accomplishments in space.

b. The U.S. has greater resources than the USSR for attaining space leadership but has failed to make the necessary hard decisions and to marshal those resources to achieve such leadership.

c. This country should be realistic and recognize that other nations, regardless of their appreciation of our idealistic values, will tend to align themselves with the country which they believe will be the world leader—the winner in the long run. Dramatic accomplishments in space are being increasingly identified as a major indicator of world leadership.

d. The U.S. can, if it will, firm up its objectives and employ its resources with a reasonable chance of attaining world leadership in space during this decade. This will be difficult but can be made probable even recognizing the head start of the Soviets and the likelihood that they will continue to move forward with impressive successes. In certain areas, such as communications, navigation, weather, and mapping, the U.S. can and should exploit its existing advance position.

e. If we do not make the strong effort now, the time will soon be reached when the margin of control over space and over men's minds through space accomplishments will have swung so far on the Russian side that we will not be able to catch up, let alone assume leadership.

f. Even in those areas in which the Soviets already have the capability to be first and are likely to improve upon such capability, the United States should make aggressive efforts as the technological gains as well as the international rewards are essential steps in eventually gaining leadership. The danger of long lags or outright omissions by this country is substantial in view of the possibility of great technological breakthroughs obtained from space exploration.

g. Manned exploration of the moon, for example, is not only an achievement with great propaganda value, but it is essential as an objective whether or not we are first in its accomplishment—and we may be able to be first. We cannot leapfrog such accomplishments, as they are essential sources of knowledge and experience for even greater successes in space. We cannot expect the Russians to transfer the benefits of their experiences or the advantages of their capabilities to us. We must do these things ourselves.

h. The American public should be given the facts as to how we stand in the space race, told of our determination to lead in that race, and advised of the importance of such leadership to our future.

i. More resources and more effort need to be put into our space program as soon as possible. We should move forward with a bold program, while at the same time taking every practical precaution for the safety of the persons actively participating in space flights.

****

As for the specific questions posed in your memorandum, the following brief answers develop from the studies made during the past few days. These conclusions are subject to expansion and more detailed examination as our survey continues.

Q.1- Do we have a chance of beating the Soviets by putting a laboratory in space, or by a trip around the moon, or by a rocket to land on the moon, or by a rocket to go to the moon and back with a man. Is there any other space program which promises dramatic results in which we could win?

A.1- The Soviets now have a rocket capability for putting a multi-manned laboratory into space and have already crash-landed a rocket on the moon. They also have the booster capability of making a soft landing on the moon with a payload of instruments, although we do not know how much preparation they have made for such a project. As for a manned trip around the moon or a safe landing and return by a man to the moon, neither the U.S. nor the USSR has such capability at this time, so far as we know. The Russians have had more experience with large boosters and with flights of dogs and man.

Hence they might be conceded a time advantage in circumnavigation of the moon and also in a manned trip to the moon. However, with a strong effort, the United States could conceivably be first in those two accomplishments by 1966 or 1967.

There are a number of programs which the United States could pursue immediately and which promise significant world-wide advantage over the Soviets. Among these are communications satellites, and navigation and mapping satellites. These are all areas in which we have already developed some competence. We have such programs and believe that the Soviets do not. Moreover, they are programs which could be made operational and effective within reasonably short periods of time and could, if properly programmed with the interests of other nations, make useful strides toward world leadership.

Q.2- How much additional would it cost?

A.2- To start upon an accelerated program with the aforementioned objectives clearly in mind, NASA has submitted an analysis indicating that about $500 million would be needed for FY 1962 over and above the amount currently requested of the Congress. A program based upon NASA's analysis would, over a ten-year period, average approximately $1 billion a year above the current estimates of the existing NASA program.

While the Department of Defense plans to make a more detailed submission to me within a few days, the Secretary has taken the position that there is a need for a strong effort to develop a large solid-propellant booster and that his Department is interested in undertaking such a project. It was understood that this would be programmed in accord with the existing arrangement for close cooperation with NASA, which Agency is undertaking some research in this field. He estimated they would need to employ approximately $50 million during FY 1962 for this work but that this could be financed through management of funds already requested in the FY 1962 budget. Future defense budgets would include requests for additional funding for this purpose; a preliminary estimate indicates that about $500 million would be needed in total.

Q.3- Are we working 24 hours a day on existing programs? If not, why not? If not, will you make recommendations to me as to how work can be speeded up?

A.3- There is not a 24-hour-a-day work schedule on existing NASA space programs except for selected areas in Project Mercury, the Saturn-C-1 booster, the Centaur engines and the final launching phases of most flight missions. They advise that their schedules have been geared to the availability of facilities and financial resources, and that hence their overtime and 3-shift arrangements exist only in those activities in which there are particular bottlenecks or which are holding up operations in other parts of the programs. For example, they have a 3-shift 7-day-week operation in certain work at Cape Canaveral; the contractor for Project Mercury has averaged a 54-hour week and employs two or three shifts in some areas; Saturn C-1 at Huntsville is working around the clock during critical test periods while the remaining work on this project averages a 47-hour week; the Centaur hydrogen engine is on a 3-shift basis in some portions of the contractor's plants.

This work can be speeded up through firm decisions to go ahead faster if accompanied by additional funds needed for the acceleration.

Q.4- In building large boosters should we put our emphasis on nuclear, chemical or liquid fuel, or a combination of these three?

A.4- It was the consensus that liquid, solid and nuclear boosters should all be accelerated. This conclusion is based not only upon the necessity for back-up methods, but also because of the advantages of the different types of boosters for different missions. A program of such emphasis would meet both so-called civilian needs and defense requirements.

Q.5- Are we making maximum effort? Are we achieving necessary results?

A.5- We are neither making maximum effort nor achieving results necessary if this country is to reach a position of leadership.

[signed]
Lyndon B. Johnson

READING NO. 11

PRESIDENT KENNEDY ANNOUNCES
THE APOLLO DECISION[1]

John F. Kennedy unveiled the commitment to execute Project Apollo before Congress on 25 May 1961 in a speech on "Urgent National Needs," billed as a second State of the Union message. In the speech he asked for support to accomplish four basic goals in space exploration, only the lunar landing is usually remembered. In addition, he asked for congressional appropriations for weather satellites, communications satellites, and the nuclear propulsion rocket. Congress agreed to all of them with barely any comment. As seen in this excerpt from the speech, Kennedy couched the space program in the context of the cold war rivalry with the Soviet Union.

γ   γ   γ

. . . Finally, if we are to win the battle that is now going on around the world between freedom and tyranny, the dramatic achievements in space which occurred in recent weeks should have made clear to us all, as did the Sputnik in 1957, the impact of this adventure on the minds of men everywhere, who are attempting to make a determination of which road they should take. Since early in my term, our efforts in space have been under review. With the advice of the Vice President, who is Chairman of the National Space Council, we have examined where we are strong and where we are not, where we may succeed and where we may not. Now it is time to take longer strides—time for this nation to take a clearly leading role in space achievement, which in many ways may hold the key to our future on earth.

[1]*Public Papers of the Presidents of the United States: John F. Kennedy, 1961* (Washington, DC: Government Printing Office, 1962), pp. 403–405.

I believe we possess all the resources and talents necessary. But the facts of the matter are that we have never made the national decisions or marshalled the national resources required for such leadership. We have never specified long-range goals on an urgent time schedule, or managed our resources and our time so as to insure their fulfillment.

Recognizing the head start obtained by the Soviets with their large rocket engines, which gives them many months of lead-time, and recognizing the likelihood that they will exploit this lead for some time to come in still more impressive successes, we nevertheless are required to make new efforts on our own. For while we cannot guarantee that we shall one day be first, we can guarantee that any failure to make this effort will make us last. We take an additional risk by making it in full view of the world, but as shown by the feat of astronaut Shepard, this very risk enhances our stature when we are successful. But this is not merely a race. Space is open to us now; and our eagerness to share its meaning is not governed by the efforts of others. We go into space because whatever mankind must undertake, free men must fully share.

I therefore ask the Congress, above and beyond the increases I have earlier requested for space activities, to provide the funds which are needed to meet the following national goals:

First, I believe that this nation should commit itself to achieving the goal, before this decade is out, of landing a man on the moon and returning him safely to the earth. No single space project in this period will be more impressive to mankind, or more important for the long-range exploration of space; and none will be so difficult or expensive to accomplish. We propose to accelerate the development of the appropriate lunar space craft. We propose to develop alternate liquid and solid fuel boosters, much larger than any now being developed, until certain which is superior. We propose additional funds for other engine development and for unmanned explorations— explorations which are particularly important for one purpose which this nation will never overlook; the survival of the man who first makes this daring flight. But in a very real sense, it will not be one man going to the moon—if we make this judgment affirmatively, it will be an entire nation. For all of us must work to put him there.

Secondly, an additional 23 million dollars, together with 7 million dollars already available, will accelerate development of the Rover nuclear rocket. This gives promise of some day providing a means for even more exciting and ambitious exploration of space, perhaps beyond the moon, perhaps to the very end of the solar system itself.

Third, an additional 50 million dollars will make the most of our present leadership, by accelerating the use of space satellites for world-wide communications.

Fourth, an additional 75 million dollars—of which 53 million dollars is for the Weather Bureau—will help give us at the earliest possible time a satellite system for world-wide weather observation.

Let it be clear—and this is a judgment which the Members of the Congress must finally make—let it be clear that I am asking the Congress and the country to accept a firm commitment to a new course of action—a course which will last for many years and carry very heavy costs: 531 million dollars in fiscal '62—and estimated seven to nine billion dollars additional over the next five years. If we are to go only half way, or reduce our sights in the face of difficulty, in my judgment it would be better not to go at all.

Now this is a choice which this country must make, and I am confident that under the leadership of the Space Committees of the Congress, and the Appropriating Committees, that you will consider the matter carefully.

It is a most important decision that we make as a nation. But all of you have lived through the last four years and have seen the significance of space and the adventure in space, and no one can predict with certainty what the ultimate meaning will be of mastery of space.

I believe we should go to the moon. But I think every citizen of this country as well as the Members of the Congress should consider the matter carefully in making their judgment, to which we have given attention over many weeks and months, because it is a heavy burden, and there is no sense in agreeing or desiring that the United States take an affirmative position in outer space, unless we are prepared to do the work and bear the burdens to make it successful. If we are not, we should decide today and this year.

This decision demands a major national commitment of scientific and technical manpower, material and facilities, and the possibility of their diversion from other important activities where they are already thinly spread. It means a degree of dedication, organization and discipline which have not always characterized our research and development efforts. It means we cannot afford undue work stoppages, inflated costs of material or talent, wasteful interagency rivalries, or a high turnover of key personnel.

New objectives and new money cannot solve these problems. They could in fact, aggravate them further—unless every scientist, every engineer, every serviceman, every technician, contractor, and civil servant gives his personal pledge that this nation of freedom, in the exciting adventure of space. . . .

READING NO. 12

THE KENNEDY ADMINISTRATION
DEFENDS APOLLO[1]

*Criticism of the priority assigned to the space program, and
particularly Project Apollo, increased in 1963. Congress tried
unsuccessfully, for example, to delete $700 million from the
NASA appropriation for Apollo. As a result, on 9 April 1963
Kennedy asked Lyndon Johnson as head of the National Aeronautics and Space Council for a careful review of the program.
Johnson replied on 13 May with a lengthy report that emphasized
the positive results of the space program and noted challenges
that it posed. In the end, as these excerpts suggest, Johnson's
report reflected the administration's continued commitment to an
aggressive lunar landing program for international prestige,
scientific, and cost benefit reasons.*

γ   γ   γ

. . . II. BENEFITS TO NATIONAL ECONOMY
FROM NASA SPACE PROGRAMS

1.   It cannot be questioned that billions of dollars directed into
research and development in an orderly and thoughtful manner
will have a significant effect upon our national economy. No
formula has been found which attributes specific dollar values
to each of the areas of anticipated developments, however, the
"multiplier" of space research and development will augment
our economic strength, our peaceful posture, and our standard
of living.
2.   Even though specific dollar values cannot be set for these
benefits, a mere listing of the fields which will be affected is

[1]Lyndon B. Johnson to the President, 13 May 1963, with attached report, John F.
Kennedy Presidential Files, NASA Historical Reference Collection, NASA
Headquarters, Washington, DC.

convincing evidence that the benefits will be substantial. The
benefits include:

    a. Additional knowledge about the earth and the Sun's
influence on the earth, the nature of interplanetary space
environment, and the origin of the solar system as well as of
life itself.

    b. Increased ability and experience in managing major re-
search and development efforts, expansion of capital facilities, encouragement of higher standards of quality production,

    c. Accelerated use of liquid oxygen in steelmaking, coat-
ings for temperature control of housing, efficient transfer of
chemical energy into electrical energy, and wide-range advances in electronics.

    d. Development of effective filters against detergents; in-
creased accuracy (and therefore reduced costs) in measuring
hot steel rods; improved medical equipment in human care;
stimulation of the use of fiberglass refractory welding tape,
high energy metal forming processes; development of new
coatings for plywood and furniture; use of frangible tube
energy absorption systems that can be adapted to absorbing
shocks of failing elevators and emergency aircraft landings.

    e. Improved communications, improved weather forecast-
ing, improved forest fire detection, and improved navigation.

    f. Development of high temperature gas-cooled graphite
moderated reactors and liquid metal cooled reactors; development of radioisotope power sources for both military and
civilian uses; development of instruments for monitoring
degrees of radiation; and application of thermoelectric and
thermionic conversion of heat to electric energy.

    g. Improvements in metals, alloys, and ceramics.

    h. An augmentation of the supply of highly trained techni-
cal manpower.

    i. Greater strength for the educational system both through
direct grants, facilities and scholarships and through setting
goals that will encourage young people.

    j. An expansion of the base for peaceful cooperation among
nations.

k. Military competence. (It is estimated that between $600 and $675 million of NASA's FY 1964 budget would be needed for military space projects and would be budgeted by the Defense Department, if they were not already provided for in the NASA budget.)

III. PROBLEMS RESULTING
FROM THE SPACE PROGRAM

1. The introduction of a vital new element into an economy always creates new problems but, otherwise, the nation's space program creates no major complications. The program has, to a lesser magnitude, the same problems which Defense budgets and programs have been creating for several years.

2. Despite claims to the contrary, there is no solid evidence that research and development in industry is suffering significantly from a diversion of technical manpower to the space program. NASA estimates that:

a. The nation's pool of scientists and engineers was 1,400,000 as of January 1, 1963.

b. NASA programs employed 42,000 of these scientists and engineers— only 9,000 directly on NASA payrolls.

c. On this basis, the NASA space program currently draws upon only 3% of the national pool of scientists and engineers.

d. Taking into account anticipated expansion, NASA programs are not expected to absorb more than 7% of our country's total supply of scientists and engineers.

3. The majority of the technical people working for NASA fall in the category of engineering. However, NASA's education programs are designed to help the universities train additions to the nation's technical manpower needs.

4. NASA has undertaken to support the annual graduate training of 1000 Ph.D.s, 1/4 of the estimated overall shortage of 4,000 per year. This program would more than replace those drawn upon by the agency.

5. In overall terms, NASA finds that diversion of manpower and resources is not a major problem arising from the space program. A major problem, however, is the need to minimize waste

and inefficiency. To help meet this challenge, turnover of top level Government talent should be reduced and compensation more in line with responsibilities would contribute to this objective. . . .

CONCLUSION

There is one further point to be borne in mind. The space program is not solely a question of prestige, of advancing scientific knowledge, of economic benefit or of military development, although all of these factors are involved. Basically, a much more fundamental issue is at stake—whether a dimension that can well dominate history for the next few centuries will be devoted to the social system of freedom or controlled by the social system of communism.

The United States has made clear that it does not seek to "dominate" space and, in fact, has led the way in securing international cooperation in this field. But we cannot close our eyes as to what would happen if we permitted totalitarian systems to dominate the environment of the earth itself. For this reason our space program has an overriding urgency that cannot be calculated solely in terms of industrial, scientific, or military development. The future of society is at stake.

READING NO. 13

POST-APOLLO SPACE
PROGRAM PLANNING[1]

In 1963 the officials in the Kennedy administration began to consider what program NASA should undertake after the completion of Apollo. After Kennedy's assassination in November 1963, his successor asked NASA Administrator James E. Webb to identify future objectives for the civilian space program. Webb was quite reluctant to commit NASA to specific goals and priorities in advance of any expression of political support, preferring instead to list a range of possible tasks and to ask top policymakers to choose the options they wished to pursue. This was the approach taken in this January 1965 report by NASA to President Johnson. It provided an overview of the capabilities NASA was developing and the uses to which they might be applied. As the introduction and summary of this report reproduced here demonstrate, Webb refused to priorize and make recommendations except to continue a "balanced" space effort.

γ   γ   γ

I. INTRODUCTION
The successful flight of Sputnik I, in its most fundamental aspect, meant that man had taken the first step toward the exploration of a new environment by means of a new technology. It also meant that in the USSR, which accomplished this first step, new horizons were opened and there was a surge of national pride and accomplishment. An internal drive was created that changed the posture of Soviet society and lifted it above many of the frictions and tensions of the existing status.

---

[1]NASA, "Summary Report: Future Programs Task Group," January 1965, Program Planning Files, NASA Historical Reference Collection, NASA Headquarters, Washington, DC.

Horizons were widened. Internationally, the leadership and image of the Soviet Union were vastly enhanced. The flights of Gagarin and other Soviet Cosmonauts added impetus to a marked degree.

In the United States and in the Free World, as we all know, the immediate effects were quite the opposite. However, since then, we have made tremendous progress under a broad based and balanced program aimed at achieving pre-eminence in aeronautics and space.

Down through the course of history, the mastery of a new environment, or of a major new technology, or of the combination of the two as we now see in space, has had profound effects on the future of nations; on their relative strength and security; on the relations with one another; on their internal economic, social and political affairs; and on the concepts of reality held by their people. . . .

The long-range effects of man's entry into space, in person and by instruments and machines, can be best forecast in terms of these considerations. As a new environment, space may well become as important to national security and national development as the land the oceans and the atmosphere; rockets and spacecraft as important as automobiles, trucks, trains, ships, submarines and aircraft. The foreseeable returns from scientific advances, technical advances, and practical uses compare favorably with the returns yielded by the most vigorous past periods of exploration of newly opened segments of man's expanding frontier.

If these larger considerations of the space effort are to be adequately dealt with in terms of national policy, they must be translated into broad objectives in order that particular programs and missions can be defined and evaluated. For the United States, these objectives relate aeronautics to space and are contained in the Space Act of 1958. . . . Under the Space Act, NASA bears the general responsibility for continuously providing an adequate underlying aeronautical and space capability and cooperating with the military services and other agencies which have, or anticipate, specific missions and uses. In 1958, and again in 1961, two major periods of wide debate and assessment brought decisions to undertake missions and programs which accelerated our progress toward the achievement of these objectives. The capability which has been created through the work thus begun and now under way will be the basis for this analysis.

First, however, we need to understand that we face certain conditions and constraints.

II. SUMMARY OF CONDITIONS AND CONSTRAINTS FOR FUTURE PLANNING

In planning future missions in space, there are a number of conditions and constraints which must be considered. These are:

a. The space activity of the Soviet Union indicates a vigorous program in near-Earth, lunar, planetary, and manned space fields. The Soviet announced program is still broader, and extends to communications and meteorology. The Russians have exploited their initial advantage in launch vehicle power, have upgraded that power, and have shown great skill in exercising their systems. The United States must, of course, meet its own needs in space rather than accept the judgment of another nation. We must decide the pace at which our needs will be met. Our decisions must be related to other national needs and priorities and we must recognize that at this early stage in space exploration the full meaning of space cannot be forecast. Still, Soviet actions must be taken into account as indicating values that they see and seek. Their skill in exploiting space spectaculars to their advantage in the areas of national prestige, and international politics must be recognized and countered. At a minimum, the United States must achieve a basic knowledge of space environments and systems, and must maintain an operational capability sufficient to guarantee full access to and use of space.

b. As a result of the work of the past 6 years, we already have a broad base of scientific and technical knowledge about many factors of space and are entering a period of rapidly expanding launch capability which we will use to achieve a far broader scientific and technical base and to gain wide experience in manned space flight. Our seven successful Saturn I test launches point to the operational use of the much greater boost power of the Saturn IB by 1966 and to the use of the giant Saturn V booster by 1967. The Defense Department's Titan Ill-C booster is expected to begin flight testing in the near future. The application of Centaur as an upper stage can augment considerably the capabilities of the larger vehicles.

In the critical area of determining man's capabilities in space, Gemini operations with two-man crews, supported by a world-wide net of stations and recovery forces and managed from our new mission control center at Houston, will begin in 1965. Hundreds of man hours of flight experience with mission durations up to 14 days will provide experience in rendezvous, docking, manned performance outside the spacecraft and other operations, adding large increments to our knowledge. Apollo operations with three-man crews beginning in 1967 will significantly extend the Gemini experience so that by the time we undertake the first manned landing on the Moon, the Gemini and Apollo programs will have provided thousands of man hours of flight time. Additional general and specialized manned flight experience is expected from the Manned Orbiting Laboratory (MOL) under development by the Department of Defense.

In science, the measurements and knowledge acquired over the next few years will become increasingly valuable as they are used and refined. More sophisticated questions will be asked by scientists and more sophisticated spacecraft, such as the Orbiting Astronomical Observatory (OAO) and the Surveyor, together with manned spacecraft, will be used to search far answers. In communications and meteorology, the imminent operational use of satellite systems will answer many questions as to the current value of these systems. Further research will answer many questions as to their potential for the future.

These activities will uncover new uses of space, force the solution of many emerging technical problems, and reveal others not now identified. This has been the way of all new areas of scientific and technical development, and it will certainly be true in space.

It is, therefore, essential that in making plans for the future all elements required for a balanced program and the practice of maintaining maximum flexibility be preserved.

c. The United States decided in 1958, and reaffirmed the decision in 1961, to move rapidly in all areas of space exploration. The wide range of power in the boosters in our National Launch Vehicle Program and the wide scope of completed and on-going missions reflects this fundamental decision. Our missions extend from sounding rocket explora- tion of the near-Earth atmosphere; to orbiting solar, geophysi- cal and astronomical observatories; to manned exploration of the Moon; and to probes out to Venus and Mars. New capabilities, both in science and in the practical uses of space, have been developed and these capabilities are being rapidly expanded.

As we emerge from this initial period of space exploration, we can see that our 1958 decision to mount a broad, balanced program was a wise one. The scientific and technical returns to date clearly indicate the value of continuing within this framework. In areas where a definitive point has not yet been reached, as in manned flight, we face the urgent need to push forward to such a point.

d. Of major importance to future planning is the fact that, in NASA, approved programs are making heavy demands on limited financial and human resources. At present budget levels, this situation will continue for at least another year. Manned flights of Gemini will begin shortly; the Saturn IB must be completed and mated to the Apollo spacecraft for manned flights in earth orbit; the Saturn V must be brought along and everything possible done to achieve a manned lunar mission in this decade. In addition, both the Lunar Orbiter and Surveyor programs must be pushed vigorously and in proper relation to the Apollo requirements in design, produc- tion and flight.

Unless the NASA budget is to be increased, the present situation requires that priority be given in the near-term period to insuring that these programs proceed without delay, under vigorous budgetary control, and produce successful results. As this is accomplished, we can, within present budgetary levels, begin to make plans to move toward other large goals in space and undertake new missions. The Saturn IB/Centaur booster can provide the power for a large Voyager mission to Mars by 1971. The Apollo system will gain in reliability through repeated use over the next several years and will then provide much of the launch and spacecraft capability to meet expanded goals.

Therefore, unless an urgent National need arises, large new mission commitments can, better than in previous periods, be deferred for further study and analysis based heavily on ongoing advanced technological developments and flight experience.

e. During the past 6 years, the United States, in addition to carrying out many specific missions, has been engaged in creating a large, basic space capability. This includes a versatile family of launch vehicles and spacecraft, and the trained manpower and facilities to continue to produce and use them. Other things being equal, these capabilities, with the resultant gain in reliability, should be used to the maximum degree in future missions. If such capabilities are not used they deteriorate. Their use can achieve lower costs and greater assurance of success for many desirable future programs than can otherwise be achieved.

In summary, the NASA manned program for the future, and the developments leading toward it, . . . require that we must:

First, apply available resources to every aspect required for success in the on-going programs, especially the Apollo program, and to bring these to fruition as quickly and efficiently as possible.

Second, define an intermediate group of missions and work toward them using the capability being created in the on-going programs. The launch vehicles and spacecraft being developed in the Apollo program are of such size, versatility and efficiency as to be of decisive importance in achieving and maintaining pre-eminence in space during the next period. Significant use of this capability over and above the presently programmed Apollo mission can be made as early as Calendar Year 1968, if small increments of additional resources are committed soon.

As now planned, by 1969 the Apollo program will build up to capability to launch six Saturn IB systems and six Saturn V systems annually. The Saturn IB will boast 35,000 pounds or more to Earth orbit and, when mated to the Centaur, can send up to 10,000 pounds of unmanned payload to Mars, of which up to

5,000 pounds can be landed an the planet. The Saturn V will place 250,000 pounds or more in Earth orbit or send 95,000 pounds to the vicinity of the Moon. These powerful boosters, in these numbers, will make it possible for the first time for this Nation to regularly launch large payloads in operational space systems far a wide variety of purposes, in manned and unmanned flights in the Earth-Moon region, and for unmanned planetary missions.

Further, the present program will provide the industrial base and operational capability for producing and using eight Apollo-LEM spacecraft systems annually. A broad spectrum of manned space flight missions can be carried out with these systems. In near-Earth space, missions can include low inclination, polar, or synchronous orbits to accomplish scientific, technological, and applications objectives. Such missions can utilize extensive maneuvering and extra-vehicular operations. Such flexible multi-purpose use of the Apollo-LEM system can be accomplished by reduction of crew size and utilization of payload margins either for expendable supplies or propellants. It is entirely feasible to extend the manned Earth orbital stay time of the Apollo-LEM system to 30 days and possibly to as much 90 days. Similarly, the Apollo-LEM system capability for lunar missions can be extended to permit detailed mapping of the Moon from lunar polar orbit; and the Lunar Excursion Module can be used as a truck to provide supplies for longer stay time and increased exploration capability on the lunar surface. For unmanned planetary exploration, the Saturn IB, using the Centaur as a third stage, can not only launch a 10,000-pound Voyager spacecraft with a 5,000-pound lander to Mars for the 1971 opportunity, but can meet the increased volume and power requirements for later years.

If one or more such missions are selected, intensive study should begin immediately and preparations be made for undertaking firm commitments for procurement of additional launch vehicles and spacecraft about July 1966. Program definition of the Saturn IB-Centaur mating should begin immediately, but major flight hardware expenditures need not begin until 1966. With respect to modification of spacecraft life support and power systems far extended manned operations, the fabrication of components required for specific missions would not be required until 1966, but supporting activity, including advanced development, should be expanded above the present level of effort.

Third, continue long range planning of missions that might be initiated late in this decade or early in the 1970's. Appropriate research and development work to provide the science and technology for these missions should be carried forward. Such long range plans could include, for example, such manned areas as an Earth-orbiting space station, a lunar base with roving vehicle, and a planetary spacecraft.

This suggested pattern for future planning is reflected in the following detailed presentation of (1) present capabilities; (2) intermediate missions; and (3) long-range missions. It should be noted that the boundary between categories (2) and (3) is subject to constant reevaluation and possible revision in the light of increased knowledge and changing events. . . .

VI. SUMMARY

Our study of future programs has covered three major categories . . .

a. A review of the capabilities being developed by current programs;

b. Intermediate missions which would support National objectives in space and afford steady progress toward longer-range goals, and at the some time make most effective use of capabilities developed thus far; and

c. Long-range missions which may comprise the Nation's space exploration goals in the decades ahead.

In the areas of aeronautics, satellite applications, unmanned and manned space exploration, launch vehicles, and research and technology development, it is possible to trace horizontally the development path from 1958 to a decade or further into the future. It is obvious that there is increased uncertainty as the plans are projected into the future.

The details of these new missions, such as specific spacecraft designs and exact mission plans will, of course, be the subject of continued study by Headquarters and Field Center of NASA, by interested government agencies, by universities, and by industry. Continued space exploration will be an evolutionary process in which the next step is based largely on what was learned from the experience of preceding research and flight missions. The pace at which these new programs will be carried out will necessarily depend upon many other factors, such as the allocation of budgetary and manpower resources and the changing National needs of the future.

This study has not revealed any single area of space development which appears to require an overriding emphasis or a crash effort. Rather, it appears that a continued balanced program, steadily pursuing continued advancement in aeronautics, space sciences, manned space flight, and lunar and planetary exploration, adequately supported by a broad basic research and technology development program, still represents the wisest course. Further, it is believed that such a balanced program will not impose unreasonably large demands upon the Nation's resources and that such a program will lad to a pre-eminent role in aeronautics and space.

READING NO. 14

SPACE SCIENCE AND APPLICATIONS MANAGEMENT[1]

When James Webb took charge of NASA, he instituted regular program reviews. Once each month, on Saturday mornings, one of the program offices would review its efforts. In June 1967, Homer Newell and his senior staff presented their philosophy for management of NASA's space science and applications program. The excerpts from this document provide an excellent snapshot of the space science program in 1967 and describe program philosophy and content, the project review process, and program problems and their solutions.

γ   γ   γ

. . . The objectives of OSSA [Office of Space Science and Applications] management are to achieve in an efficient, effective, and acceptable manner various objectives of the national Space Program. The *objects* of management include: technical problems, schedules, resources, facilities, people, institutions. The *tools* of management include: funding and other resources, personnel, organization, and a variety of processes and procedures.

Management is *efficient* if it conserves resources—including time—handles problems in an orderly fashion rather than on a crisis-to-crisis basis, exhibits a reasonable durability of management solutions, and evidences proper foresight of future developments. Management is *effective* if it accomplishes established objectives within applied resource constraints. Management is *acceptable* if it strengthens rather than weakens those

---

[1] NASA, Office of Space Science and Applications, "Program Review: Science and Applications Management," 22 June 1967, Space Science and Applications Files, NASA Historical Reference Collection, NASA Headquarters, Washington, DC.

with whom it deals—including ourselves, our sister offices within NASA, and agencies and institutions outside of NASA.

A management pattern should be an appropriate function of time. While a good management arrangement should exhibit substantial durability, there should also be an orderly mechanism for making changes to meet changing needs. Caution should be exercised to avoid becoming too enamored of a chosen scheme, or too comfortable with the status quo. The science and applications management pattern that we will discuss did not "just happen," but is the product of careful and serious thought on the part of many, . . . The OSSA management pattern is reviewed continually, and periodically is subjected to detailed scrutiny in order to maintain the proper continuing evolution in the context of changing needs. . . .

The Scientific Community

The viability of our Space Science Program depends on the competence and quality of the scientists working with us. It is our conviction that NASA must have highly competent scientific groups within its Centers. At the same time, the majority of our space science effort is, and we believe should be, carried out by scientists outside of NASA, primarily in the university community. We draw the scientific community into our program through the opportunities to do research on space missions. We have also drawn leading scientists into our planning effort through the subcommittees to our Space Science Steering Committee—and through association with the National Academy of Sciences (especially the Space Science Board), the President's Science Advisory Committee, and other similar organizations. Special ad hoc studies, like the two Space Science Summer Studies conducted in the past, and the Applications Summer Study in prospect, add to the scope of our planning and expand our contact with the outside community.

The emergence of large-scale scientific projects, like Voyager and the creation of a large astronomical facility in orbit, appear to call for new arrangements for continuing our association with the scientific community. We are, as you know, working on this problem. As one step in the direction of organizing for the future, we have recently created the Lunar and Planetary Missions Board under the chairmanship of Dr. John Findlay. At the same time, we are discussing with leading members of the astronomical community appropriate arrangements for the future in the astronomy area.

The Universities

A large part of the outside scientific community resides in our universities and colleges. In carrying out our Space Science and Applications Programs we have, therefore, developed an extensive association with the university community. Past associations have been very productive and rewarding, and we will want to continue many of them in much the same form as in the past. On the other hand, just as the changing character of our program calls for changes in our relations with individual scientists, so also we will probably need to develop new types of association with the universities as institutions. At the present time there is under discussion the possibility of creating special institutes in areas such as lunar exploration and space astronomy. We are examining possibilities such as these with the help of the National Academy of Sciences. These matters are not simple. It would be very easy to create something undesirable to our long-lasting regret. The problem before us is to work with the Academy and the universities to strengthen all of us as we work together in carrying out the national space program. We in OSSA welcome the opportunity to work with the new University Affairs Office on problems such as these.

Relations with Other Government Agencies

OSSA has numerous relations with other Government agencies, such as the Departments of Defense, Commerce, Transportation, Interior, and Agriculture, and the Atomic Energy Commission. There is no need to elaborate upon these relationships in this brief review. I would, however, like to single out one important point for attention.

Many of the users of the space technology and capabilities that we are developing within NASA are to be found in our sister government agencies. The Department of Commerce has responsibilities and interests in meteorology, the space environment, and oceanography. The Departments of Interior and Agriculture have special interests in the earth resources area. The Federal Aviation Agency of the Department of Transportation is concerned with navigation and air traffic control and safety. The Federal Communications Commission has special interest in our communications developments. The Department of Defense has a wide ranging interest in those technologies that can contribute to military and national defense applications.

Our Applications Office must, therefore, develop a close rapport with these various agencies both in the planning and in the conduct of space programs designed ultimately to support the agencies in meeting important national needs. The molding of such relationships can become very complex, as in the case of the Communications Satellite Corporation, where both government and commercial interests become mixed in an arena where national policy has yet to be defined completely.

The Aerospace Industry

Over 82 percent of the Science and Applications R&D dollar is expended with industry in the development of hardware. The majority (58%) of this total is concentrated in 21 major contractors. Problems associated with schedule, cost and performance are primarily evident during the development phase, and we have experienced most of these problems in one or more of the several developments which have been successfully completed. We have worked closely with our industrial partners in developing science and applications projects, and through this relationship, seek to contribute to the national capability for accomplishing complex technological undertakings.

As a source of applied research, the aerospace industry bears to our Applications Program the sort of relationship that the scientific and university basic research communities bear to our Space Science Program. In the Applications Program, we already have a strong relationship with industry through the contract work on applications missions, and through special contracted studies. We are endeavoring to draw applications-oriented and engineering communities more into the total planning activity than has been true in the past. One of the means for this will be the Applications Summer Study, to be carried out by the National Academies of Science and Engineering during the next two summers. . . .

Space Science Program Management

The latest count shows 942 scientists from 297 institutions involved as investigators in our program. We flew 121 experiments on spacecraft in 1966. We also flew 99 sounding rockets. We are currently funding 1,822 individual tasks. . . . There are 216 scientists involved as consultants in 16 committees, boards, and subcommittees which report to you but work through our office—such as the Heyns' Ad Hoc Science Advisory Committee and the Lunar and Planetary Missions Board.

Annually we send to about 5,500 scientists a document announcing opportunities to participate in the NASA space program. (In addition, 800 copies of this document were sold by the Government Printing Office.) In 1966 we evaluated 366 proposals for flight experiments, 248 of which were selected for flight. An additional 1,329 unsolicited proposals for SR&T work were evaluated. This whole operation is supervised from Headquarters by about 20 key scientists and engineers.

In addition to the number and variety of tasks, management must take into account the undisciplined nature of the scientists and scientific activity. If you asked a typical academic scientist what his management philosophy is you are likely to get the reply, "Neither a manager nor managed be." However, the chaotic nature of scientific activity is not a problem unique to NASA—many other agencies, such as ONR, NSF, and NIH, have wrestled with that problem. What is unique to NASA and what constitutes our most complex management problem is the coupling of the undisciplined scientific activity into a highly disciplined engineering and administrative activity—the design, preparation, and conduct of a space mission.

With these thoughts in mind, let's consider how we accomplish this coupling. . . . One of our major objectives is a sound research program. To achieve this we try to select from among the many possible research programs those which appear most likely to produce significant discoveries. When we consider a potential space science mission we first ask, "Are the scientific objectives worthwhile?" If the answer is "yes" then we ask, "Is it technically feasible?"; "Are there sufficient people to do it?"; "Can we get the funds to support it?" A sound research program is difficult to achieve because it is hard to predict the best lines of research to pursue. However, there is one thing we can be sure of—we won't have a sound research program unless we have good scientists deeply involved in all phases of the program. The administrative procedures we use were developed to involve good scientists in the program.

This brings me to the next two major objectives—the efficient use of resources and the maintenance of a scientific and technical base. The time and talents of the 942 scientists involved in the program are valuable national resources which we must use wisely. If they are participating in a sound research program we are a long way toward achieving that goal. However, we must maintain the proper balance between the numbers of scientists involved in the program and the number of flight opportunities. If we have many more scientists developing flight experiments than there are flight opportunities, then the scientists will not be able to do sufficient research to have a productive career and we will be wasting a national resource and jeopardizing these people's careers.

On the other hand, if we have many more flight opportunities than sound experiments to perform, we start to fill our spacecraft with second-rate experiments, which is inefficient use of the funds and facilities available.

Why is the maintenance of a scientific and technical base one of our management objectives? Why can't we contract with industry for a collection of instruments, put these on the shelf until we have a flight ready, fly them, and make the data available to the scientific community? We can't because research is a continuing, evolving process. If a scientist is going to make a significant discovery in a field, he must be working in that field; he must know what the latest results are; he must know the best instrument to use. In short, intellectually, he must be committed to the field. Experiments purchased and put on the shelf likely will be out of date when flown. Consequently, the data will be of little interest to a good scientist.

An international program enables us to get more research done because of the contributions of the other country. In some cases of global phenomena, the only way to do the research is through an international program. There is an additional very real value of the international program in that the scientists and engineers who participate in the program develop an increased understanding and respect for our people and our institutions.

A rapid and accurate dissemination of the knowledge gained is vital to the full value of the space program.

We should consciously experiment with new procedures and institutions for the management of science. The unique problems of the management of space research may force us to

develop methods which will be valuable to other agencies
dealing with research. The experimentation which led to our
present procedures was done in the late fifties and early sixties.
We have not had to change them much since then. However, we
are now entering a new era where we must develop the procedures and institutions to cope with Voyager, with the permanent
astronomical facilities, and with manned exploration of the
Moon.

All of these objectives support the central objective of maintaining U.S. achievement and leadership in science and applications. . . .

A major cornerstone of our science policy is the performance
of the major share of space science research by academic
scientists. It is essential that university scientists be deeply
involved in the space program because a large portion of the best
scientists are at universities. The university environment attracts
the creative scientists, and thus, by involving them in the
program, we tap a source of good people and good ideas. We
also generate new scientists and engineers interested in and able
to conduct space research. Finally, through publications, books,
and teaching, the academic scientist is in an admirable position
to disseminate the knowledge acquired.

Another cornerstone of our policy is in-house scientific competence at those Centers heavily involved in the management of
scientific missions. Good scientists can be attracted to NASA
Centers by the facilities that are available and by the opportunity
to influence in major ways the shaping of the space science
program. In addition to providing creative research programs,
these scientists are a source of new concepts for spacecraft
systems and missions. It is good far the morale of the Center to
have some of its own scientists conducting experiments on
spacecraft flown by the Center. The in-house scientist also plays
another very important role. He serves as the communication
channel between the Center Project Manager and the academic
scientist. Any lingering doubts we may have had about the need
for in-house scientific competence have been dispelled over past
years when we have tried to conduct space science programs
from Centers which did not have a strong in-house competence.

Our basic philosophy can be summarized by saying that
we concentrate the scientific competence in three places—

universities, NASA Centers, and NASA Headquarters. Scientists at universities and NASA Centers compete for flight opportunities by submitting proposals whose scientific merit is judged by their scientific peers. Competent, mature scientists at Headquarters establish, defend, and conduct a national program which provides equal opportunity for any scientist to participate. . . .

Very early in the history of NASA it became obvious that the selection of scientific payloads for our missions was one of the most important decisions to be made. The quality of the scientific payload determines the quality of the results which enter the scientific literature and upon which history ultimately will judge our performance. Not only was the selection of a good payload a necessity to meet the scientific objectives, but it was also fraught with all sorts of human implications. The selection of a scientist for a particular flight could establish his scientific reputation and take another perhaps equally good person out of the running.

Accordingly, the decision was made that payloads would be selected by Headquarters, specifically by Dr. Newell, who was then Deputy Director of Space Flight Programs. To assist him in the selection of payloads, the Space Science Steering Committee was formed. Although the membership of the Steering Committee has enlarged and changed membership, the basic pattern of having each program office represented by its top scientist and its top engineer has not changed. The center slide shows the present makeup of the Steering Committee.

Soon after the Committee began operating, it became clear that more scientific talent was needed to provide the proper technical review. Accordingly, several discipline subcommittees were formed . . ., chaired by the appropriate Headquarters Program Chief, with the most competent scientists in the discipline as members, whether they were at a NASA Center or at universities. The pattern that has emerged is a Subcommittee of about 12–17 members, with 5–9 of these being academic scientists and the remainder being from NASA Centers, other Government laboratories, and industry. . . .

As payloads became more complex, it became obvious that the Center project organizations must be involved in the selection process. Dr. Newell could select a fine scientific payload only to find after sending it to the Center that only half of the experiments were compatible with the spacecraft. Accordingly, another step in the selection process was added, and the proposed payload is sent to the Center to review the experiments for compatibility with the spacecraft and to review the investigator's organization to see if he will be able to produce flightworthy hardware on schedule. . . .

Although this process may seem long and cumbersome, you will recall from my first figure that in 1966 some 366 flight proposals went through the system, from which 248 flight experiments were selected. Furthermore, the system selects good experiments which make the flight schedule and produce reliable data during flight. This system has also produced a large number of scientists capable of producing flight hardware. For example, in the area of particles and field in 1960, there were about 20 capable experimenters in that discipline, while today there are almost 80.

The results from the 985 experiments chosen by this procedure since its inception in 1960 have established the United States as the clear leader in space science. I know of no better testimony than that to the validity of the process. . . .

READING NO. 15

NASA DECIDES TO MAKE A CIRCUMLUNAR APOLLO FLIGHT[1]

In the aftermath of the tragic Apollo 204 capsule fire in 1967, NASA's goal of reaching the Moon before the end of the decade seemed in jeopardy. It took almost twenty months after the fire, until October 1968, before astronauts were launched into orbit aboard an Apollo spacecraft. The success of this test flight, however, prompted the Apollo program manager, Air Force General Samuel C. Phillips, to suggest a bold strategy for regaining momentum in the Lunar landing program. He recommended in November 1968 that the next Apollo flight be recast as a circumlunar mission. His memorandum, accepted by the NASA administrator on 18 November 1968, made possible the dramatic mission of Apollo 8 on 21–27 December 1968.

γ   γ   γ

11 Nov 1968

The purpose of this memorandum is to obtain your approval to fly Apollo 8 on an open-ended lunar orbit mission in December 1968.

My recommendation is based on an exhaustive review of pertinent technical and operational factors and also on careful consideration of the impact that either a success or a failure in this mission will have on our ability to carry out the manned lunar landing in 1969. . . .

On August 19, 1968, we announced the decision to fly Apollo 8 as a Saturn V, CSM [Command/Service Module]-only mis-

---

[1]NASA, Apollo Program Director, to NASA, Associate Administrator for Manned Space Flight, "Apollo 8 Mission Selection," 11 November 1968, Apollo 8 Files, NASA Historical Reference Collection, NASA Headquarters, Washington, DC.

sion. The basic plan provided for Apollo 8 to fly a low earth orbital mission, but forward alternatives were to be considered up to and including a lunar orbital mission. Final decision was to be reserved pending completion of the Apollo 7 mission and a series of detailed reviews of all elements of the Apollo 8 mission including the space vehicle, launch complex, operational support system, and mission planning. . . .

The decision process, resulting in my recommendation, had included a comprehensive series of reviews conducted over the past several weeks to examine in detail all facets of the considerations involved in planning for and providing a capability to fly Apollo 8 on a lunar orbit mission. . . .

PROS AND CONS OF A LUNAR ORBITAL FLIGHT:

My objective through this period has been to bring into meaningful perspective the trade-offs between total program risk and gain resulting from introduction of a CSM-only lunar orbit mission on Apollo 8 into the total mission sequence leading to the earliest possible successful Apollo lunar landing and return. As you know, this assessment process is inherently judgmental in nature. Many factors have been considered, the evaluation of which supports a recommendation to proceed forward with an Apollo 8 open-ended lunar orbit mission. These factors are:

PROS:

Mission Readiness:
—The CSM has been designed and developed to perform a lunar orbit mission and has performed very well on four unmanned and one manned flights . . .
- We have learned all that we need in earth orbital operation except repetition of performance already demonstrated.
- The extensive qualification and endurance-type subsystem ground testing conducted over the past 18 months on the CSM equipments has contributed to a high level of system maturity, as demonstrated by the Apollo 7 flight.
- Performance of Apollo 7 systems has been thoroughly reviewed, and no indication has been evidenced of design deficiency.
- Detailed analysis of Apollo 4 and Apollo 6 launch vehicle anomalies, followed by design modifications and rigorous

ground testing gives us high confidence in successful performance of the Apollo 8 launch vehicle.
- By design all subsystems affecting crew survival . . . are redundant and can suffer significant degradation without crew or mission loss. . . .
- Excellent consumables and performance margins exist for the first CSM lunar mission because of the reduction in performance requirements represented by omitting the weight of the lunar module. . . .

Effect on Program Progress:
The lunar orbit mission will:
- Provide valuable operational experience on a lunar CSM mission for flight and ground and recovery crews. This will enhance probability of success on the subsequent more complex lunar missions by permitting training emphasis on phases of these missions as yet untried.
- Provide an opportunity to evaluate the quality of MSFN and on-board navigation in lunar orbit including the effects of local orbit perturbations. This will increase anticipated accuracy of rendezvous maneuvers and lunar touchdown on a lunar landing mission.
- Permit validation of Apollo CSM communications and navigation systems at lunar distance.
- Serve to improve consumables requirements prediction techniques.
- Complete the final verification of the ground support elements and the onboard computer programs.
- Increase the depth of understanding of thermal conditions in deep space and lunar proximity.
- Confirm the astronauts' ability to see, use, and photograph landmarks during lunar orbit.
- Provide an early opportunity for additional photographs for operational and scientific uses such as augmenting Lunar Orbiter coverage and for obtaining data for training crewmen on terrain identification under different lighting conditions.

CONS:

Mission Readiness:

- Marginal design conditions in the Block II CSM may not have been uncovered with only one manned flight.
- The life of the crew depends on the successful operation of the Service Propulsion System during the Transearth Injection maneuver.
- The three days endurance level required of backup systems in the event of an abort from a lunar orbit mission is greater than from an earth orbit mission.

CONS:

Effect on Program Progress:
- Validation of Colossus spacecraft software program and Real Time Computer Complex ground software program could be accomplished in a high earth orbital mission.
- Only landmark sightings and lunar navigation require a lunar mission to validate.

Impact of Success or Failure on Accomplishing Lunar Landing in 1969:

A successful mission will:
- Represent a significant new international achievement in space.
- Offer flexibility to capitalize on success and advance the progress of the total program toward a lunar landing without unreasonable risk.
- Provide a significant boost to the morale of the entire Apollo program, and an impetus which must, inevitably enhance our probability of successful lunar landing in 1969.

A mission failure will:
- Delay ultimate accomplishment of the lunar landing mission.
- Provide program critics an opportunity to denounce the Apollo 8 mission as precipitous and unconservative.

RECOMMENDATION:
In conclusion, but with the proviso that all open work against the Apollo 8 open-ended lunar orbit mission is completed and certified, I request your approval to proceed with the implementation plan required to support an earliest December 21, 1968, launch readiness date.

READING NO. 16

NEIL ARMSTRONG SETS FOOT ON THE MOON[1]

After eight years of all-out effort, nearly $20 billion expended, and three astronauts' deaths, on 20 July 1969 Apollo 11 landed on the Moon. The two astronauts who set foot on the surface, Neil A. Armstrong and Edwin E. Aldrin, called it in what later astronauts thought of as an understatement, "magnificent desolation." This document contains the radio transmissions of the landing and Armstrong's first venture out onto the Lunar surface. The "CC" in the transcript is Houston Mission Control, CDR is Neil Armstrong, and LMP is Buzz Aldrin.

γ   γ   γ

04 06 45 52	CC	We copy you down, Eagle [the name of the Lunar Module].
04 06 45 57	CDR	Houston, Tranquility Base here.
04 06 45 59	CDR	THE EAGLE HAS LANDED.
04 06 46 04	CC	Roger, Tranquility. We copy you on the ground. You got a bunch of guys about to turn blue. We're breathing again. Thanks a lot.
04 06 46 16	CDR	Thank you.
04 06 46 18	CC	You're looking good here.
04 06 46 23	CDR	Okay. We're going to be busy for a minute. . . .
04 06 46 38	LMP	Very smooth touchdown. . . .
04 06 47 03	LMP	Okay. It looks like we're venting the oxidizer now. . . .

[1]NASA, Manned Spacecraft Center, "Apollo 11 Technical Air-to-Ground Voice Transcription," July 1969, pp. 317, 377, Apollo 11 Files, NASA Historical Reference Collection, NASA Headquarters, Washington, DC.

04 06 47 09 CC   Eagle you are STAY for T1 [one day on Moon].

04 06 47 12 CDR   Roger. Understand, STAY for T1. . . .

04 13 23 38 CDR   [After suiting up and exiting the Lunar Module (LM), Armstrong was ready to descend to the Moon's surface]. I'm at the foot of the ladder. The LM footpads are only depressed in the surface about 1 or 2 inches, although the surface appears to be very, very fine grained, as you get close to it. It's almost like a powder. Down there, it's very fine.

04 13 23 43 CDR   I'm going to step off the LM now.

04 13 24 48 CDR   THAT'S ONE SMALL STEP FOR MAN, ONE GIANT LEAP FOR MANKIND.

04 13 24 48 CDR   And the—the surface is fine and powdery. I can—I can kick it up loosely with my toe. It does adhere in fine layers like powdered charcoal to the sole and sides of my boots. I only go in a small fraction of an inch, maybe an eighth of an inch, but I can see the footprints of my boots and the treads in the fine, sandy particles.

04 13 25 30 CC   Neil, this is Houston. We're copying.

04 13 25 45 CDR   There seems to be no difficulty in moving around as we suspected. It's even perhaps easier than the simulations at one-sixth g that we performed in the various simulations on the ground. It's actually no trouble to walk around. Okay. The descent engine did not leave a crater of any size. It has about 1 foot clearance on the ground. We're essentially on a very level place here. I can see some evidence of rays emanating from the descent engine, but a very insignificant amount. . . .

READING NO. 17

NASA PLANS A POST-APOLLO
SPACE PROGRAM[1]

In part to buy time and avoid making a firm decision that might be premature, soon after entering the presidential office in January 1969, Richard Nixon appointed a Space Task Group to study post-Apollo plans and make recommendations. Chartered on 13 February 1969 under the chairmanship of Vice President Spiro T. Agnew, this group met throughout the spring and summer to plot a course for the space program. NASA lobbied hard with the Group and especially its chair for a far-reaching post-Apollo space program that included a mission to Mars, a space station, and a reusable Space Shuttle. This call for an aggressive civil space program was reflected in the group's report on 15 September 1969, as this excerpt demonstrates.

γ   γ   γ

The next two decades of space activities can make significant progress in scientific returns, applications to human needs, and exploration of the solar system. The critical factors that will make efficient progress possible are:

- -Commonality—the use of a few major systems for a wide variety of missions.
- -Reusability—the use of the same system over a long period for a number of missions.
- -Economy—the reduction in the number of "throw away" elements in any mission; the reduction in the number of new developments required; the development of new program principles that capitalize on such capabilities as man-tending of space facilities; and the commitment to simplification of space hardware.

---

[1]NASA's Report to the President's Space Task Group, *The Post-Apollo Space Program: Directions for the Future* (Washington, DC: Executive Office of the President, September 1969), pp. 1–2.

Central to the program of the next 20 years is the continuing extension of man's abilities into space. This generalized capability can be achieved through a sufficiently large, multi-manned long-lived space station module. In order to support the station and its subsequent additions, efficient transportation to and from Earth is required. Low operating costs and high flexibility are necessary here to take full advantage of the manned orbital station's unique capabilities for scientific, engineering, and applications functions. Therefore, an Earth-to-orbit shuttle must be developed that can meet these criteria; this envisages high reusability and eventually, airline-type operations between space and the surface of the Earth. Together, the space station and shuttle are the keystones to the next major accomplishments of the nation's space program.

The space station and shuttle's relationship to future mission possibilities are of basic importance:

-By assembling a number of individual station modules together in orbit, a permanent space base of large size can be launched with the existing Saturn launch vehicles. Shuttles would provide rapid easy access by and support to crews, experimenters, and operators.

-Later, a space station in polar orbit around the Moon using the same modules would provide a flexible base from which to conduct stage, or space tug, capable of long staytime on the lunar surface with 3- to 6-man teams.

-This same tug could land complete space station modules on the Moon as the building blocks of a permanent lunar surface base. Resupply from Earth would be most economically achieved through use of a reusable Earth orbit-to-lunar-orbit transportation system, a restartable and space storable nuclear stage capable of making many round trips between Earth and Moon. Crews, payload, and propellants would be carried to Earth orbit by the shuttle. This system will greatly lower costs of establishing and supporting lunar bases.

-Eventual manned expeditions to Mars in the 1980's would employ the basic space station module as crew quarters and laboratory while in transit and as an orbital base while at the planet. These expeditions would use the same nuclear stages as would the lunar bases. A new excursion module to land men on Mars from orbit would be the only major new development required.

Three representative program approaches for the next 20 years have been developed, each of which maintains the integrity of these basic concepts:

- Increasing returns in science, engineering, and applications commensurate with investment costs.
- Pursuit of the principles of commonality, reusability, and economy through the efficient development and management of manned space systems.
- Continued exploration of the solar system, including manned expeditions to the planets, beginning with Mars.

The three programs include the same principal elements of activity but the accomplishments are differently phased in time with a corresponding shift in annual resource requirements.

READING NO. 18

NIXON'S RESPONSE TO THE SPACE
TASK GROUP REPORT[1]

Richard Nixon did not react to the Space Task Group report until 7 March 1970, and when he did so he called for a much less ambitious effort. This short document summarizes the space policy of the United States through virtually all of the 1970s, a period when NASA formulated a space program that tried to accomplish all of the goals of this statement in an austere budget environment.

γ   γ   γ

Over the last decade, the principal goal of our nation's space program has been the Moon. By the end of that decade men from our planet had traveled to the Moon on four occasions and twice they had walked on its surface. With these unforgettable experience[s], we have gained a new perspective of ourselves and our world.

I believe these accomplishments should help us gain a new perspective of our space program as well. Having completed that long stride into the future which has been our objective for the past decade, we must now define new goals which make sense for the Seventies. We must build on the successes of the past, always reaching out for new achievements. But we must also recognize that many critical problems here on this planet make high priority demands on our attention and our resources. By no means should we allow our space program to stagnate. But—with the entire future and the entire universe before us—we should not try to do everything at once. Our approach to space must continue to be bold—but it must also be balanced.

---

[1]White House Press Secretary, "The White House, Statement by the President," 7 March 1970, Richard M. Nixon Presidential Files, NASA Historical Reference Collection, NASA Headquarters, Washington, DC.

When this Administration came into office, there were no clear, comprehensive plans for our space program after the first Apollo landing. To help remedy this situation, I established in February of 1969 a Space Task Group, headed by the Vice President, to study possibilities for the future of that program. Their report was presented to me in September. After reviewing that report and considering our national priorities, I have reached a number of conclusions concerning the future pace and direction of the nation's space efforts. The budget recommendations which I have sent to the Congress for Fiscal Year 1971 are based on these conclusions.

Three General Purposes

In my judgment, three general purposes should guide our space program.

One purpose is exploration. From time immemorial, man has insisted on venturing into the unknown despite his inability to predict precisely the value of any given exploration. He has been willing to take risks, willing to be surprised, willing to adapt to new experiences. Man has come to feel that such quests are worthwhile in and of themselves—for they represent one way in which he expands his vision and expresses the human spirit. A great nation must always be an exploring nation if it wishes to remain great.

A second purpose of our space program is scientific knowledge—a greater systematic understanding about ourselves and our universe. With each of our space ventures, man's total information about nature has been dramatically expanded; the human race was able to learn more about the Moon and Mars in a few hours last summer than had been learned in all the centuries that had gone before. The people who perform this important work are not only those who walk in spacesuits while millions watch or those who launch powerful rockets in a burst of flame. Much of our scientific progress comes in laboratories and offices, where dedicated, inquiring men and women decipher new facts and add them to old ones in ways which reveal new truths. The abilities of these scientists constitute one of our most valuable national resources. I believe that our space program should help these people in their work and should be attentive to their suggestions.

A third purpose of the United States space effort is that of practical application turning the lessons we learn in space to the early benefit of life on Earth. Examples of such lessons are manifold; they range from new medical insights to new methods of communication, from better weather forecasts to new management techniques and new ways of providing energy. But these lessons will not apply themselves; we must make a concerted effort to see that the results of our space research are used to the maximum advantage of the human community.

A Continuing Process

We must see our space effort, then, not only as an adventure of today but also as an investment in tomorrow. We did not go to the Moon merely for the sport of it. To be sure, those undertakings have provided an exciting adventure for all mankind and we are proud that it was our nation that met this challenge. But the most important thing about man's first footsteps on the Moon is what they promise for the future.

We must realize that space activities will be a part of our lives for the rest of time. We must think of them as part of a continuing process—one which will go on day in and day out, year in and year out—and not as a series of separate leaps, each requiring a massive concentration of energy and will and accomplished on a crash timetable. Our space program should not be planned in a rigid manner, decade by decade, but on a continuing flexible basis, one which takes into account our changing needs and our expanding knowledge.

We must also realize that space expenditures must take their proper place within a rigorous system of national priorities. What we do in space from here on in must become a normal and regular part of our national life and must therefore be planned in conjunction with all of the other undertakings which are also important to us. The space budget which I have sent to Congress for Fiscal Year 1971 is lower than the budget for Fiscal Year 1970, a condition which reflects the fiscal constraints under which we presently operate and the competing demands of other programs. I am confident, however, that the funding I have proposed will allow our space program to make steady and impressive progress.

Six Specific Objectives

With these general considerations in mind, I have concluded that our space program should work toward the following specific objectives:

1. We should continue to *explore the Moon*. Future Apollo manned lunar landings will be spaced so as to maximize our scientific return from each mission, always providing, of course, for the safety of those who undertake these ventures. Our decisions about manned and unmanned lunar voyages beyond the Apollo program will be based on the results of these missions.

2. We should move ahead with bold exploration of the planets and the universe. In the next few years, scientific satellites of many types will be launched into Earth orbit to bring us new information about the universe, the solar system, and even our own planet. During the next decade, we will also launch unmanned spacecraft to all the planets of our solar system, including an unmanned vehicle which will be sent to land on Mars and to investigate its surface. In the late 1970s, the "Grand Tour" missions will study the mysterious outer planets of the solar system—Jupiter, Saturn, Uranus, Neptune, and Pluto. The positions of the planets at that time will give us a unique opportunity to launch missions which can visit several of them on a single flight of over three billion miles. Preparations for this program will begin in 1972.

There is one major but long range goal we should keep in mind as we proceed with our exploration of the planets. As a part of this program we will eventually send men to explore the planet Mars.

3. We should work to *reduce substantially the cost of space operations*. Our present rocket technology will provide a reliable launch capability for some time. But as we build for the longer-range future, we must devise less costly and less complicated ways of transporting payloads into space. Such a capability—designed so that it will be suitable for a wide range of scientific, defense, and commercial uses—can help us real-

ize important economies in all aspects of our space program. We are currently examining in greater detail the feasibility of reusable space shuttles as one way of achieving this objective.

4.   We should seek to *extend man's capability to live and work in space*. The Experimental Space Station (XSS)—a large orbiting workshop—will be an important part of this effort. We are now building such a station—using systems originally developed for the Apollo program—and plan to begin using it for operational missions in the next few years. We expect that men will be working in space for months at a time during the coming decade.

We have much to learn about what man can and cannot do in space. On the basis of our experience with the XSS, we will decide when and how to develop longer lived space stations. Flexible, long-lived space station modules, could provide a multi-purpose space platform for the longer-range future and ultimately become a building block for manned interplanetary travel.

5.   We should *hasten and expand the practical applications of space technology*. The development of earth resources satellites—platforms which can help in such varied tasks as surveying crops, locating mineral deposits and measuring water resources—will enable us to assess our environment and use our resources more effectively. We should continue to pursue other applications of space-related technology in a wide variety of fields, including meteorology, communications, navigation, air traffic control, education, and national defense. The very act of reaching into space can help man improve the quality of life on Earth.

6.   We should *encourage greater international cooperation in space*. In my address to the United Nations last September, I indicated that the United States will take positive, concrete steps "toward internationalizing man's epic venture into space—an adventure that belongs not to one nation but to all mankind." I believe that both the adventures and the applications of space missions should be shared by all peoples. Our progress will be faster and our accomplishments will be greater if nations will join together in this effort, both in contributing the resources and in enjoying the benefits. Unmanned scientific payloads from other nations already make use of our space launch capability on

a cost-shared basis; we look forward to the day when these arrangements can be extended to larger applications satellites and astronaut crews. The Administrator of NASA recently met with the space authorities of Western Europe, Canada, Japan and Australia in an effort to find ways in which we can cooperate more effectively in space.

* * *

It is important, I believe, that the space program of the United States meet these six major objectives. A program which achieves these goals will be a balanced space program, one which will extend our capabilities and knowledge and one which will put our new learning to work for the immediate benefit of all people.

As we enter a new decade, we are conscious of the fact that man is also entering a new historic era. For the first time, he has reached beyond his planet; for the rest of time, we will think of ourselves as men *from* the planet Earth. it is my hope that as we go forward with our space program, we can plan and work in a way which makes us proud *both* of the planet from which we come *and* of our ability to travel beyond it.

READING NO. 19

A PLEA FOR THE SPACE SHUTTLE[1]

When Richard Nixon refused to embrace the overall Space Task Group report, NASA began looking for a way to gain approval of at least a portion of the program outlined. Its leaders recognized that a space station, lunar colony, and mission to Mars were the most ambitious parts of the long-range space plan and could not be funded in the tight fiscal environment of the early 1970s, but that the Space Shuttle had some potential because it could stand alone and had a relatively well-defined price tag. Even so, obtaining presidential approval was difficult. A breakthrough came on 12 August 1971 when Casper W. Weinberger, Deputy Director of the Office of Management and Budget (OMB), wrote a memorandum to the president, arguing that "there is real merit to the future of NASA, and to its proposed programs" and suggesting that Nixon approve the start of Space Shuttle development. In a handwritten scrawl on Weinberger's memo, Richard Nixon indicated "I agree with Cap." The memorandum set the stage for a formal announcement by President Nixon on 5 January 1975 to build the Space Shuttle.

γ   γ   γ

Present tentative plans call for major reductions or change in NASA, by eliminating the last two Apollo flights (16 and 17), and eliminating or sharply reducing the balance of the Manned Space Program (Skylab and Space Shuttle) and many remaining NASA programs.

[1]Caspar W. Weinberger memorandum to the President, via George Shultz, "Future of NASA," 12 August 1971, White House, Richard M. Nixon, President, 1968–1971 File, NASA Historical Reference Collection, NASA Headquarters, Washington, DC.

I believe this would be a mistake.

1) The real reason for sharp reductions in the NASA budget is that NASA is entirely in the 28% of the budget that is controllable. In short we cut it because it is cuttable, not because it is doing a bad job or an unnecessary one.

2) We are being driven, by the uncontrollable items, to spend more and more on programs that offer no real hope for the future: Model Cities, OEO, Welfare, interest on National Debt, unemployment compensation, Medicare, etc. Of course, some of these have to be continued, in one form or another, but essentially they are programs, not of our choice, designed to repair mistakes of the past, not of our making.

3) We do need to reduce the budget, in my opinion, but we should not make all our reduction decisions on the basis of what is reducible, rather than on the merits of individual programs.

4) There is real merit to the future of NASA, and to its proposed programs. The Space Shuttle and NERVA particularly offer the opportunity, among other things, to secure substantial scientific fall-out for the civilian economy at the same time that large numbers of valuable (and hard-to-employ-elsewhere) scientists and technicians are kept at work on projects that increase our knowledge of space, our ability to develop for lower cost space exploration, travel, and to secure, through NERVA, twice the existing propulsion efficiency of our rockets.

It is very difficult to re-assemble the NASA teams should it be decided later, after major stoppages, to re-start some of the long-range programs.

5) Recent Apollo flights have been very successful from all points of view. Most important is the fact that they give the American people a much needed lift in spirit, (and the people of the world an equally needed look at American superiority). Announcement now, or very shortly, that we were cancelling Apollo 16 and 17 (an announcement we would have to make very soon if any real savings are to be realized) would have a very bad effect, coming so soon after Apollo 15's triumph. It would be confirming in some respects, a belief that I fear is gaining credence at home and abroad: That our best years are behind us, that we are turning inward, reducing our defense

commitments, and voluntarily starting to give up our super-power status, and our desire to maintain world superiority.

America should be able to afford something besides increased welfare, programs to repair our cities, or Appalachian relief and the like.

6) I do not propose that we necessarily fund all NASA seeks—only that . . . we *are* going to fund space shuttles, NERVA, or other major, future NASA activities. . . .

READING NO. 20

NASA VERSUS THE SPACE SCIENCE COMMUNITY[1]

This memo demonstrates something of the "love-hate" relationship between NASA and the science community that began almost with the creation of NASA and has persisted to the present. The fundamental issues have not changed; they only grew more complex. Neither NASA nor the Space Science Board understood the difficulties faced by the other, and both organizations struggled to control the space science agenda.

γ   γ   γ

Following are some points to keep in mind when discussing the matter of NASA-Space Science Board relationships. . . . As we wrestle with the problem of maintaining good relations with the outside science community, we [can]not forget that we have good scientists within NASA who also need to be heard, and who need to be assured of the opportunity to derive professional satisfaction from their work. . . .

First of all some background. Relations with the Space Science Board, and also with our own Boards and Committees, began to come apart about the time the Space Task Group Report was published. Strains developed because the Boards and Committees felt they were not being effective or listened to by NASA. The budgets in the Space Task Group Report were regarded as appallingly high. The emphasis given to large-scale programs—space shuttle, space stations, space bases, lunar bases, nuclear shuttles, Grand Tours, and manned missions to

[1]Homer E. Newell, NASA Associate Administrator for Space Science and Applications, to James C. Fletcher, NASA Administrator, "Relations with the Science Community and the Space Science Board," 3 December 1971, James C. Fletcher Chronological File, 1971, NASA Historical Reference Collection, NASA Headquarters, Washington, DC.

Mars—had a very negative effect. Our own Lunar and Planetary Missions Board threatened to resign en masse. The Lunar
and Planetary Missions Board took strong exception to the order
of choice of some of the near-term planetary missions, but their
more serious concern was that if they had been aware of the total
context developed in the Space Task Group Report, the entire
scope of their advice on planetary program would have been
different. It was this kind of concern, which was also expressed
by the Astronomy Missions Board, that led us to form the Space
Program Advisory Council with Chairmen of our advisory
Committees as members, so that in the Council these Chairmen
can get the kind of perspective that was not available to our
former Boards.

It was in this period of turmoil and reaction to the kind of
program proposed in the Space Task Group Report, that disen-
chantment with Viking and active concern over large scale
planetary projects like Grand Tour began to develop. When
Harry Hess died, and Herb Friedman took over as temporary
Chairman of the Space Science Board, these difficulties were
exacerbated. Friedman has long been a strong proponent of the
smaller types of space projects, and very much wants to see the
sounding rocket work enlarged, particularly astronomy. Additionally, Dr. Friedman is personally intensely interested in high-
energy and infrared astronomy, and wishes to see them pushed,
if necessary, at the expense of ultraviolet and other areas of
astronomy. This hostility to the large-scale projects, including
particularly the large-scale manned space flight programs proposed for the future, on the part of the Acting Chairman of the
Space Science Board, naturally made itself felt, and carried
over into the handling of the Woods Hole Summer Study of
1970, which Dr. Friedman chaired.

Charlie Towns, very much at our urging and the Academy's,
took over chairmanship of the Space Science Board at the time
that we in NASA were planning a restructuring of our advisory
committees in order to eliminate weaknesses experienced in the
earlier arrangement. When he took over, Charlie undertook to
revamp the Board, instituting a policy of rotation in membership. Thus many new members were brought on, and many of
the older members rotated off. As a consequence the Space
Science Board, like our space program advisory council and its

Committees, is going through something of a learning period. Having more continuity than SPAC and its Committees, the Space Science Board has less of a problem at present in this regard than do our Council and Committees, but we should also recognize that the small size of the Board means that some areas of space science may not enjoy the same degree of representation or advocacy as others. Thus some divergence of scientific opinion must be expected from time to time.

The Space Science Board has for the past year and a half clearly been feeling its way toward what it would like to regard as the proper relationship with NASA. The recent meeting at the Jet Propulsion Laboratory was something of a milestone in this regard, when we went through in detail for the Board the various problems and considerations involved in our FY 1973 budget decisions, and outlined in as much detail as was available various options for the future in planetary exploration. A large number of the Board members indicated satisfaction with that meeting. As Roman Smoluchowski put it, prior to the JPL meeting the relationship between the Board and NASA had been largely that of adversaries, while as the result of the meeting the relationship seemed more like that of partners. This brings me to the first major point that I think ought to be established between us and the Space Science Board.

1. There is a need for more exposure on both sides to the give and take of problems and alternatives being considered on the other side. To continue the development of a feeling of partnership between NASA and the Board, we in NASA need to give more attention to ensuring that the Board has a good insight into the budget problems, political pressures, technical tradeoffs, manpower restraints, etc., that we are wrestling with. The Board, on the other hand, needs to be more open with us. There is a great tendency nowadays for the Board to call executive sessions, excluding even senior NASA personnel from them, to debate points of view, pros and cons for different alternatives, opinions regarding NASA plans and approaches, etc., and then after a position has been determined to present that final position to NASA people. In the days of Berkner and Harry Hess, Dryden and I were welcomed into such discussion meetings, where we heard all the give and take, and were consequently more sensitive and alert

to the feelings of the Board and its members on objectives, relative priorities, approaches, etc. In those days I detected no hesitation on the part of the Board members to express concern or dissatisfaction over things that NASA was doing, and the ensuing discussions often led to constructive approaches to resolving problems. . . .

2. A second important way to enhance the feeling that we are working together is to involve at least the chairman of the Space Science Board, and perhaps the Chairman plus some selected members, in close discussions with Naugle and the Administrator during those last weeks of decisions when our budget proposal is taking final shape. Such discussions will enable us to detect whether misunderstandings of Space Science Board positions or intensity of feeling about different alternatives are leading to decisions that might otherwise not be made. Such discussions, though informal and off the record, would help reassure the Board that its views are indeed being taken into consideration in the final decisions. Then, even if a decision goes contrary to or differs in some respect from a Space Science Board recommendation, at least there will be an understanding of the reasons, and a better awareness of the fact that the decision was not an arbitrary rejection of the Board's position.

3. The Board has felt of late that NASA tends to provide backup support—studies, analysis, budget estimates, etc.— for those missions or projects that NASA wants to consider, but not adequate support for those that the Space Science Board would like to consider. We should spend more time discussing with the Board the kinds of studies and analysis that they need in order to carry out their advisory role. This will, of course, take manpower and dollars, but a modest investment in this area should pay big dividends.

4. Finally, we need to develop some way of joining hands in support of long-term projects, so that programs that we start with the support and encouragement of the Space Science Board don't later founder because the support is withdrawn or watered down midway through the effort. This should be helped by a fuller and more open discussion between NASA and the Board in early phases of planning. NASA needs to be sensitive to the concerns which may appear mild in the early

phases, but which may grow in intensity as the project proceeds until they become a serious stumbling block. The Space Science Board, on the other hand, must avoid underplaying concerns that are in their view serious. The Board also should be urged, as was done by you at JPL, to discipline itself to regard a recommendation to undertake a project as a commitment by the Board to stand by and work with NASA not only during the early days of selling the project to the Administration and the Congress, but also during the long period of carrying it out. With rotating membership the Board will need to develop some sense of continuity of opinion, so the one year's membership understand and support the recommendations made by a different group of people some years earlier.

The form of our relationship with the Space Science Board and the Scientific community will be of no real significance unless it also produces a program that the Board and Scientific community can believe in and support. In this regard, there are a few points that the Space Science Board has made over and over for many years. Some of these are summarized in the next paragraph.

5. For a long time the Board has consistently urged that NASA develop a balanced program, balanced not only with respect to different disciplines, but also balanced within each discipline between large and small projects. An essential element of such a balanced program is, in the view of the Board, the flexibility and quick turn-around time afforded by smaller projects and smaller spacecraft. Short lead times, and the ability to follow up quickly new scientific discoveries, are viewed as essential to a good scientific program. To achieve this flexibility and followup capability, the Board has repeatedly recommended more sounding rockets, Explorers and Pioneers, for example. The serious concern of the Board over large projects like Viking and the Grand Tour is two-fold. First at some dollar level such projects become far more expensive than they feel they can in good conscience justify on scientific grounds alone; and, secondly, through their incessant demands for funds and manpower the large projects tend to squeeze out the essential smaller projects. Moreover, with the smaller projects in the total program mix, the larger

projects make their best contribution and can be accepted as scientifically fruitful and worthwhile; without the proper number and kinds of the smaller projects, the larger projects are regarded as not producing the best science for the money invested.

6. Over the years, the Space Science Board and our own committees have expressed deep concern about the long term involved in space missions, which in some cases take between five and ten years for accomplishment. If, as space experiments become more sophisticated and more complex, they also take more and more time, then this time becomes a substantial fraction of a man's career, more of a fraction than most scientists would like to gamble on a program as risky as is space experimenting. This again argues for an adequate number of smaller projects and experiments that scientists can use to generate a steady flow of results. But equally important is the fact that if a long period intervenes between the selection of the experiment and the time of its accomplishment, this virtually guarantees that the experiment will be out of date or not the best that could be done at the time it is performed. As a consequence the SSB urges every effort on NASA's part to shorten lead times for experiments, and to make it possible to update or replace experiments as close to launch time as can be managed. One suggestion is that a group of scientists be formed for each major mission to keep in touch with the payload and advise on its status, recommending on desirable update and improvement during the course of the project. In response to these recommendations, John Naugle suggests that for each project we consciously allow for some contingency in both payload weight capacity and project dollars to accommodate new experiments late in the game, and that we ask Fowler's Physical Science Committee to review periodically the project status and make recommendations. (This is a very difficult problem. Naugle is personally committed to doing as much in this direction as possible. However, it must be pointed out that it just does take a long time to prepare these experiments for flight, and that the longest lead times are usually associated with scientific experiments, not with the spacecraft and housekeeping hardware.)

It is very natural for those who are in the middle of carrying out space projects, and of planning for the future, should have the broadest range of alternatives in mind and have strong feelings about what can and can't be accomplished, and what ought to be undertaken next. This sometimes results in NASA's moving out on a project before the scientific community has a full appreciation of what is involved. When this happens, NASA leadership often has to work on the problem of organizing its following right at the time when strong support from the outside may be crucial to sustaining Administration and Congressional support for the project. My last point refers to this problem.

7. We need to restore a NASA posture of being urged by the scientific community to do things, rather than urged not to do them. In many cases this may well be a matter of how we work new projects into the program. To be specific, a possible program for the next decade is that of building a large astronomical orbiting facility for use by the astronomical community. In his recent oral report to you, Jesse Greenstein indicated that his Academy study committee has given the large space telescope an important place on the committee's list of priorities. Our Astronomy Missions Board regarded this as an important project and recommended that we move in that direction. Recently, Herb Friedman spoke to me urging that we use the large space telescope as the principal focus for our manned space flight program beyond Apollo. Thus, support for the building of a large orbiting space telescope is beginning to form. At this point, however, NASA could easily outrun its support and generate some undesirable resistance. It might be well at this stage to move more slowly, consolidating interest in and support for such astronomy projects step by step, until we arrive at a stage where pressure for the large space telescope is so great that we can hardly fail to accede to it. . . .

READING NO. 21

NIXON APPROVES THE SPACE SHUTTLE[1]

On 11 December 1971 NASA Administrator James C. Fletcher met with presidential assistant Peter M. Flanigan, science advisor Edward E. David, Jr., and representatives of the Office of Management and Budget (OMB) to discuss the Shuttle program. He learned at that meeting that Nixon had decided in principle to go ahead with the project, but that some additional decisions over size and cost had yet to be made. On 3 January 1972 Fletcher and Low met with Caspar Weinberger of OMB and learned of the final decision to build a partially reusable configuration with specific cargo capacity that would meet both defense and NASA specifications for $5.15 billion. Fletcher flew to San Clemente, California, for a meeting with the president on 5 January, to announce the decision. The announcement printed here brought relief both to the aerospace industry and to space advocates. Both ballyhooed it as a great step forward in national capability. Critics derided it as an ill-timed, ill-considered, unnecessary expenditure of public funds.

γ γ γ

I have decided today that the United States should proceed at once with the development of an entirely new type of space transportation system designed to help transform the space frontier of the 1970s into familiar territory, easily accessible for human endeavor in the 1980s and '90s.

This system will center on a space vehicle that can shuttle repeatedly from earth to orbit and back. It will revolutionize transportation into near space, by routinizing it. It will take the astronomical costs out of astronautics. In short, it will go a long

[1] White House Press Secretary, "The White House, Statement by the President," 5 January 1972, Richard M. Nixon Presidential Files, NASA Historical Reference Collection, NASA Headquarters, Washington, DC.

way toward delivering the rich benefits of practical space utilization and the valuable spinoffs from space efforts into the daily lives of Americans and all people.

The new year 1972 is a year of conclusion for America's current series of manned flights to the moon. Much is expected from the two remaining Apollo missions—in fact, their scientific results should exceed the return from all the earlier flights together. Thus they will place a fitting capstone on this vastly successful undertaking. But they also bring us to an important decision point—a point of assessing what our space horizons are as Apollo ends, and of determining where we go from here.

In the scientific arena, the past decade of experience has taught us that spacecraft are an irreplaceable tool for learning about our near-earth space environment, the moon, and the planets, besides being an important aid to our studies of the sun and stars. In utilizing space to meet needs on earth, we have seen the tremendous potential of satellites for intercontinental communications and world-wide weather forecasting. We are gaining the capability to use satellites as tools in global monitoring and management of natural resources, in agricultural applications, and in pollution control. We can foresee their use in guiding airliners across the oceans and in bringing televised education to wide areas of the world.

However, all these possibilities, and countless others with direct and dramatic bearing on human betterment, can never be more than fractionally realized so long as every single trip from earth to orbit remains a matter of special effort and staggering expense. This is why commitment to the space shuttle program is the right next step for America to take, in moving out from our present beachhead in the sky to achieve a real working presence in space—because the space shuttle will give us routine access to space by sharply reducing costs in dollars and preparation time.

The new system will differ radically from all existing booster systems, in that most of this new system will be recovered and used again and again—up to 100 times. The resulting economies may bring operating costs down as low as one-tenth of those for present launch vehicles.

The resulting changes in modes of flight and re-entry will make the ride safer and less demanding for the passengers, so that men and women with work to do in space can "commute" aloft, without having to spend years in training for the skills and rigors of old style space flight. As scientists and technicians are actually able to accompany their instruments into space, limiting boundaries between our manned and unmanned space programs will disappear. Development of new space applications will be able to proceed much faster. Repair or servicing of satellites in space will become possible, as will delivery of valuable payloads from orbit back to earth.

The general reliability and versatility which the shuttle offers seems likely to establish it quickly as the workhorse of our whole space effort, taking the place of all present launch vehicles except the very smallest and very largest.

NASA and many aerospace companies have carried out extensive design studies for the shuttle. Congress has reviewed and approved this effort. Preparation is now sufficient for us to commence the actual work of construction with full confidence of success. In order to minimize technical and economic risks, the space agency will continue to take a cautious evolutionary approach in the development of this new system. Even so, by moving ahead at this time, we can have the shuttle in manned flight by 1978, and operational a short time later.

It is also significant that this major new national enterprise will engage the best efforts of thousands of highly skilled workers and hundreds of contractor firms over the next several years. The amazing "technology explosion" that has swept this country in the years since we ventured into space should remind us that robust activity in the aerospace industry is healthy for everyone—not just in jobs and income, but in the extension of our capabilities in every direction. The continued preeminence of America and American industry in the aerospace field will be an important part of the shuttle's "payload."

Views of the earth from space have shown us how small and fragile our home planet truly is. We are learning the imperatives of universal brotherhood and global ecology—learning to think and act as guardians of one tiny blue and green island in the trackless oceans of the universe. This new program will give more people more access to the liberating perspectives of space, even as it extends our ability to cope with physical challenges of

earth and broadens our opportunities for international cooperation in low-cost, multi-purpose space missions.

"We must sail sometimes with the wind and sometimes
against it," said Oliver Wendell Holmes, "but we must sail, and
not drift, nor lie at anchor." So with man's epic voyage into
space—a voyage the United States of America has led and still
shall lead.

READING NO. 22

PROBLEMS AND OPPORTUNITIES
AT NASA[1]

*In the aftermath of the lunar landings in December 1972,
NASA's continued viability as an institution was uncertain. In
this 9 May 1977 memorandum, James C. Fletcher, who headed
NASA from May 1971 to March 1977, reflects for his successor
Robert Frosch on the major institutional and programmatic
issues facing the agency. Of particular interest are Fletcher's
observations on keeping the NASA institutional base intact or at
least ensuring a flow of new people into the agency. The project
called LACIE (Large Area Crop Inventory Experiment) was an
Earth observation project using LANDSAT satellites. The Jupiter Orbiter Project (JOP) was later renamed Galileo.*

γ   γ   γ

Continuing our discussion in writing on some of the things
that are less sensitive, let me raise some issues not in any
particular order but simply for the record. . . .

1. Applications Program. In my view, the Applications Program is the "wave of the future" as far as NASA's public image
is concerned. It is the most popular program (other than aeronautics) in the Congress and as you begin to visit with community leaders, you will understand it is clearly the most popular
program with them as well. The Application Program consists
mainly of communications satellites, weather satellites, LACIE,
and earthquake research. There are problems in each of these
areas:

    a. *Communication Satellites.* We temporarily phased out
of this program in 1973 due to a severe budget cut. At the

[1] James C. Fletcher to Robert Frosch, "Problems and Opportunities at NASA," 9
May 1977, James C. Fletcher Chronological Files, 1977, NASA Historical
Reference Collection, NASA Headquarters, Washington, DC.

time, it seemed like industry was picking it up most rapidly and was something they could do without much help from NASA. I had serious misgivings when this decision was made since I realized that it was the part of the Applications Program which had the greatest public visibility and was the most obvious example of transfer to industry. We were able to keep a skeleton group aboard to support OTP [Office of Technology Policy] and FCC [Federal Communication Commission] and, to a limited extent, the existing ATS/CTS satellites. However, at this point in time, I believe we need to get back into the business one way or another. The search and rescue satellite was a small attempt in this direction. Also, the work we are doing with NOAA [National Oceanigraphic and Atmospheric Administration] and the Coast Guard to monitor fishing vessels within the 200-mile limit (they install the transponders) is also a small step in that direction. The National Academy study prepared under the chairmanship of Bill Davenport was a good one, and I think it is time we started following along the tracks that they recommended. I'm afraid, however, that OMB is going to give us problems.

b. *Weather Satellites*. To many, weather satellites are mostly talk and not much show. I had been at NASA four years before I realized that NOAA was not using weather satellites at all in their weather forecasting but rather used them as backup for their forecasters and occasionally for monitoring severe storms such as tornadoes and hurricanes. Weather satellites, however, have been used extensively by the Navy and by the Air Force for overseas forecasting, I think very effectively, and just recently NOAA's Numerical Weather Service in Suitland has begun making global weather forecasts for overseas construction and a variety of military uses.

The real potential, however, of weather satellites lies in the possibility of 5-day (possibly up to 2-week) forecasts and it has only been clear in the last year or two what the technical problems really are in making such forecasts. Bob Cooper is very much aware of the problem, as is Bob Jastrow of GISS, so I won't try to elaborate further on it except to say that what is really needed is some broad-gauge scientific talent to be involved rather than the specialized, narrow scope meteorologists who have been working the problem at NOAA (and for that matter at NASA also).

c. *LACIE*. The LACIE program is not going well and OMB is very much aware of this. If this program fails, it is going to reflect on NASA's credibility in the Applications area. What is needed here also is a new approach to the problem either organizationally or by using people of different technical background. The people now involved in the program at Houston are not the most talented, and they have been doing the same thing for too many years. It has not had high-level attention at Houston because, of course, the Shuttle is their main future. It may be that the program can be handled better by simply shifting the focus from Houston to Goddard. (I recommended this two years ago but got less than an enthusiastic reception from the Office of Applications.)

d. *Earthquake Research*. I'm afraid we have no program in earthquake research, but we were able to get funds from the Congress and OMB by labeling some of our "tectonic plate motion" investigations improperly. As near as I can tell, what we are doing is scientific research only and this does not relate directly to predicting earthquakes, although admittedly it might add to the scientific base on which future earthquake prediction techniques might be predicated.

There are a lot of opportunities in Applications that we may be missing which may or may not be related directly to the programs in which we are now engaged. Electronic mail, wired suburbs, the Cooper/Augenstein Global Information System and, of course, a leadership responsibility for a national climate program are all things that Al is aware we could move into; however, it does take aggressive leadership to pursue these opportunities. We don't have that in the Applications Office itself. In fact, to pursue these new programs, it might be wisest to set up a separate office outside of Applications and leave the marketing of current programs (a, b, c, and d above) to the Applications Office.

In addition to opportunities and problems, we have personnel problems in the Office of Applications, which I'd be glad to discuss with you sometime.

2. *The MSFC [Marshall Space Flight Center] Institutional Problem*. As I indicated to you in our discussion, NASA has an

overall institutional problem which arises from the fact that we had to trim out civil service staff by almost a factor of two since its peak during the Apollo days. This has caused a number of problems that go with aging institutions generally, but our problems were accelerated because of the rapid RIFing [Reductions in Force] that went on in the late 60's and early 70's. We still have a large number of competent people at NASA, but we are not bringing in new blood either at the younger age group or at the middle age level. There are three principal reasons for this. One is that some of the glamour has worn off from the Apollo days; second, there are other interesting fields in which scientists and engineers can become involved (in my judgment none of them compete with what goes on at NASA but, of course, I'm prejudiced); and, third, new employees feel insecure knowing that the last ones hired are usually the first to leave in case of a RIF. This would be a dilemma for any agency in such a situation and even though we try diligently to protect our best people, we are still in danger of approaching mediocrity.

This is especially severe at Marshall where some of the largest cuts were made. Some time in the 1980–81 period, we face severe manpower cuts at this Center. An obvious solution would be to close the Center unless some new program came along that would keep the staff fully occupied. Because of the urgent need for the talent that Marshall has for the Space Shuttle development, we have tried to put new programs there (such as space telescope, HEAO, etc.) and have allowed them to do a considerable amount of in-house work on the Shuttle to make good use of their personnel. Closing Marshall has been on OMB's agenda ever since I came to NASA, although from time-to-time they have also suggested JPL [Jet Propulsion Laboratory], Ames, and Lewis. We have always resisted this very strongly on the basis that (a) the initial cost of replacing the facility would be very high, and (b) we couldn't afford to risk the Space Shuttle program. The real reason, however, is that there is no guarantee that by closing a Center we would be allowed to build back to the institutional base we had before the closing, and we might find ourselves in the same RIF situation but be one Center smaller. The only possible solution that I can see is to get a commitment from the President himself (the OMB Director's commitment can always be overturned) that if we do close the Center we will be allowed to build back substantially in order to bring in new personnel. Most people in Headquarters would laugh at this suggestion but I think that it is one that ought to be considered early on in your tenure. My own bias, of course, would be to try to find work to put into MSFC and use the Center as a national resource, which it indeed is, but so far efforts along these lines have not been successful.

3. *Space Science Program.* As you have already undoubtedly picked up, we are in a dilemma on space science at NASA because it seems to be strongly supported by the White House (President, Science Adviser, OMB, etc.) but poorly supported by the Congress. Congress seems to go for the Applications Program, the Aeronautics Program, and the so-called "space spectaculars" such as Apollo, Skylab, ASTP, Viking, etc. Space solar power is an excellent example of such a spectacular and is a case in point. Apparently the reason for this dilemma is that OMB feels the Applications Program should more properly be left to the user agencies or to industry, which are always slow to support new satellite programs, whereas, Congress, especially the Space Committees, doesn't care about the user agencies because that's not their responsibility. In science, however, we have a clear mandate since we are our own user but somehow Congress recognizes that science of any kind is not popular among the general public (ask Herb Rowe for polls on that subject) and although low-profile science can get through Congress fairly easily, large bites such as the space telescope, JOP and Viking Follow-on (perhaps) seem to have difficult times. I have no pat solution for this dilemma except to continue to work the problem as we have been doing.

4. *Senate Power Base.* I used to raise my eyebrows when Jim Webb talked about a "power base" in Congress. Having just come from academia, this seemed a crude way of operating; however, after being here six years, I'm beginning to see what he meant. In the Senate especially, but to some degree also in the House, there are individuals who seem to sway the rest of the body. In the House this is less clear but certainly Tip O'Neill, George Mahon and to a lesser extent, Jim Wright, could be put into this category. In past years, Wilbur Mills and Eddie Hebert served that function, but I'm not sure their successors have quite moved into such strong positions.

Incidentally, Tiger Teague has a great deal of respect in the House, and when he is willing and able to acknowledge this respect, he can be very helpful; however, in recent years his health has been a definite handicap and this is one of our problems at the moment in the House Appropriations Subcommittee. Tiger says he is not going to run again, but when he gets his lightweight leg and his spirits improve, I wouldn't bet against his running.

In the Senate, however, the situation has changed drastically. When I first heard that Senator Proxmire was going to be Chairman of our Appropriations Subcommittee, it looked like "the end of the world" until we began to work the problem. It began to be clear that the ex officio votes on the Proxmire Subcommittee by the Senate Space Committee Chairman and Majority Leader were enough to swing the rest with no problem at all. The votes typically were 8 or 9 to 1, with Senator Proxmire's being the only negative vote. The loss of Senator Moss was considerable even though he did not have the leverage that some of the other Senators had. He was the Chairman of the Democratic Caucus and the #3 democrat in the Senate and on occasion could swing a fair number of votes. Senator Goldwater, of course, was the undisputed conservative leader in the Senate and consequently both sides of the house could be swayed by him. So it was not only the loss of Senator Moss but the loss of those ex officio votes that caused us to lose leverage in the Senate. Senator Stevenson is just learning the business but I think in time he, along with his strong staff members, should be great support especially if they are able to involve Senator Magnuson in helping him to influence some of their colleagues. Meanwhile, I'm afraid we are forced into falling back on the Proxmire Subcommittee itself.

Although we have strong support in Senator Stennis and, I believe, Senator Sasser on the Democratic side and I think all four on the Republican side, this is not enough to be considered strong support in the sense of adding in programs that the House may have taken out. This latter situation occurred many times in the past through the help of Senators Moss and Goldwater, but this year we simply can't count on it. On the other hand, I think the support is strong enough so that they are not likely to make further cuts.

The one redeeming feature in the Senate reorganization is the position of Senator Cranston as Majority Whip. He is a strong space supporter in his own right but, being a California Senator, has vested interests as well. Senator McClellan also has a great deal of influence as does Senator Stennis but those are primarily on the conservative side of the House and the number of conservative democrats is becoming fewer each year. Senator Jackson, of course, is powerful as in earlier years but so far has not had any impact on NASA's programs. Senator Cannon has moved up in stature since his recent reelection, having been one of the few western democratic Senators to be returned to the Senate. I had hoped that he would end up in one way or another as Chairman of one of our committees but that was not to be. I think becoming better acquainted with Senator Cannon can be a great help both by influencing votes in the Appropriations Committee and, of course, in the Commerce Committee itself.

These are all things that must be tracked very carefully, and I'm afraid roles are changing so rapidly that I can only alert you to the problems. Pete Crow and Joe Allen, I think, understand the situation pretty well and should be able to help make the appropriate contacts. Judy Cole, if she stays, is excellent on the G-2 and has a very good working relationship with the staff of the Senate Budget Committee. Although run by Senator Muskie, it is not yet clear how much impact it will have.

5. *Aeronautics*. I won't dwell on this subject since Al Lovelace is very familiar with the problem, but simply mention that we need to revive the fundamental work that the old NACA used to do. (The Aeronautical Centers should be at least as good as NRL [Naval Research Laboratory] is to the Navy, but so far not a single member of the NASA organization has been elected to the National Academy of Science as has Herb Friedman of NRL.) It is not clear how to do this but, of course, it is related to the institutional problem of bringing in stronger scientific and creative new talent.

6. *The Shuttle Launch Phase*. Undoubtedly Al [Lovelace] must have mentioned to you that my biggest concern on the Shuttle at the moment (aside from operational costs) is the technical difficulties involved in the launch phase. As you know, Houston is the lead Center for Shuttle development and performed very well on the Apollo spacecraft and the LM [Lunar

Module], and also carrying out operations in space. They had very little to do with the development of the Saturn launch vehicle, which was done out of Huntsville. Wernher von Braun and the people he brought with him both from Germany and from within the United States had an in-house capability second to none in the world.[2]

As a result, if you look back in the records, you will find very few difficulties with the Saturn itself and, in fact, the extra weight-carrying margin of the Saturn saved the Apollo program more than once. Incidentally, neither George Low nor John Yardley has had this launch vehicle background, and so with the loss of Rocco Petrone, we have never really had anybody in Headquarters who had much experience in this area.

I guess the question is, why do I consider this different from space problems generally? It comes down to something like the following: With the spacecraft itself during its flight in space and its landing and its attitude control system and its life support systems, etc., we had the capability to build highly redundant systems. So if we ran into a problem, there was usually time to find a "workaround" and, in fact, in every case except the fire on the ground, we managed a workaround good enough to bring the astronauts back. On the other hand, look at all of the things that went wrong during the Apollo/Skylab series. If we hadn't had this redundancy, we would have lost essentially every mission. In the case of the booster, however, there was no time for any significant workarounds on the ground. There is some redundancy built in but not an excessive amount. Therefore, testing analyses and engineering intuition have been the backbone of the launch vehicle business from the days of the V-2. Clearly the combined vehicle consisting of the two solid rockets, the external tank, and the Shuttle is the most complicated launch vehicle ever built. My big concern is whether or not analysis and testing on the ground are sufficient to ensure the reliability of this phase of the flight profile or whether engineering judgment and experience which were the hallmark

---

[2]Korolov played the same role in the U.S.S.R. When he died, the Soviets were not able to make a single new launch vehicle work!

of the von Braun group aren't still necessary for a guaranteed success. So far I have discussed this with Al Lovelace and Walt Williams only. Walt, of course, has had extensive experience with Air Force launch vehicles but again nothing like the experience of the Huntsville group. Perhaps I'm overly concerned about this problem, but when you consider the value of the payload even on the first flight and the consequences of a failure, I'd have to put it as one of my high-priority items for the near term.

7. *Pet Projects*. There are a number of things which I have tried to keep going because I believed in them, but most had a low level, which you may want to discontinue:

a. *Hypersonic Transport*. I have always felt, aside from environmental problems, that it would be a rather straightforward development to build a commercial vehicle for long-distance travel (say from New York to Delhi or from New York to Bahrein), but I'm not sure that the airplane is the best way to do it. I have the uncomfortable feeling that it might be simpler to remove the energy from a returning space vehicle by means of high-drag devices such as parachutes, blunt bodies, retrorockets, etc., rather than with wings. This is heresy at NASA but you must understand that I came up through the rocket route, not the airplane route. I did, however, go to the trouble of bringing in the parachute people to see whether indeed parachutes could be built that would allow large transports to be dropped through the atmosphere in much the same way as the Apollo capsule, and it always seems to be technically feasible but on the surface more optimal than the Shuttle itself. Needless to say, I didn't want to emphasize this in the middle of the Shuttle program.

b. *Heavy Lift Launch Vehicle*. For putting large quantities of payloads in space, it seems there are better ways of doing this than the Shuttle itself since the missions are all one way and all you need to recover are pieces of the launch vehicle itself. This can be done easily with parachutes. We might easily gain a factor of 10 to 1 over the cost per pound now required by the Shuttle.

c. *Solar Sailing*. I am sure you remember with some ambivalence Dick Garwin partly for his abrasive tone but also for his tremendous creativity. In 1972, he strongly urged me to look into the possibility of using lightweight materials to "sail" around the solar system. It took four years for the NASA "system" to respond. Bruce Murray picked it up and is now running with it. In my opinion this will be the way we will move out to Mars and other planets in the future even when we decide to go there with manned missions.

d. *Personal Communication Systems*. I really do believe that some day we will want to have person-to-person communications systems, not necessarily for the wristwatch variety but at least of the pocket calculator variety in which any person can dial any other person long distance from his car, from the golf course or wherever. This, I think, is a straightforward use of a high-powered, highly directional stationary satellite. I don't believe cost tradeoffs of this system have been made, and I'm not sure how much a person would pay for such a convenience.

e. *Technology Transfer*. The early studies made on the relationship between high technology and national productivity were very exciting indeed. The whole problem is not very well understood by economists and, I do not believe, other people in government. People seem to equate high technology with new inventions or new products instead of with productivity, and the picture gets all out of proportion. Paul Kochanowski, a former Brookings Fellow at NASA, understood the problem very well and I learned what little I know from him. It does seem to me that the impact on our economy of the technology such as NASA develops and as portions of DoD develop is absolutely enormous. I therefore have encouraged further economic studies of this process but you may wish to discontinue it.

f. *Broadening NASA's Responsibility*. As Al [Lovelace] has probably indicated to you, I've always felt that NASA's managerial talents as well as some of its technical talents have been under-utilized, and we ought to move into areas that are now the responsibilities of other agencies. This is a severe bureaucratic problem, and I'm not sure you'll want to get into that but if you do, the best time to do it is during a change of Administration, as you well know, before the bureaucracy becomes firmly entrenched.

8. *Public Affairs*. During the Apollo days and before, NASA

provided an excellent public information service to the media and, generally speaking, the public was well informed about the so-called "space spectaculars." At this point in time though we need to move to a public *relations* program; that is, an aggressive program to inform the public as to how their money is being spent and what they get for it. This is a much different problem and I have asked Bob Newman and Herb Rowe to put together a program plan which presumably has been done by now but is awaiting guidance. The last session we had brought out the fact that the focus on this aggressive program ought to be on applications and spinoffs, but we really hadn't come down to the heart of the matter and that is how to have one or two simple themes which describe NASA's contributions to the nation. My own feeling is that we need outside expertise on this one and although we brought in Burson-Marsteller, a first-rate Chicago outfit, I value the advice of Jim Mortensen of Young and Rubicam much more highly. Jim is a broad, thoughtful person interested in the space program and is willing to contribute his service freely when he has the time available. Todd Groo's experience in this area is also helpful. All of these latter are more creative than Herb and I have indicated to him that I wanted all of these other men to be heavily involved in any program plans for the future. You may wish to change that.

There are other items that I could mention here and still more that I will think of before I leave, but I expect I have covered 90 percent of the biggest issues.

READING NO. 23

RONALD REAGAN ANNOUNCES
THE SPACE STATION PROGRAM[1]

On 25 January 1984 President Ronald Reagan made an Apollo-like announcement to build a Space Station within a decade. A part of the State of the Union Address before Congress, Reagan's decision came after a long internal discussion as to the viability of the station in the national space program.

γ   γ   γ

. . . Our second great goal is to build on America's pioneer spirit—[laughter]—I said something funny? [Laughter] I said America's next frontier—and that's to develop that frontier. A sparkling economy spurs initiatives, sunrise industries, and makes older ones more competitive.

Nowhere is this more important than our next frontier: space. Nowhere do we so effectively demonstrate our technological leadership and ability to make life better on Earth. The Space Age is barely a quarter of a century old. But already we've pushed civilization forward with our advances in science and technology. Opportunities and jobs will multiply as we cross new thresholds of knowledge and reach deeper into the unknown.

Our progress in space—taking giant steps for all mankind—is a tribute to American teamwork and excellence. Out finest minds in government, industry, and academia have all pulled together. And we can be proud to say: We are first; we are the best; and we are so because we're free.

America has always been greatest when we dared to be great. We can reach for greatness again. We can follow our dreams to distant stars, living and working in space for peaceful, eco-

[1]*Public Papers of the Presidents of the United States: Ronald Reagan, 1984* (Washington, DC: Government Printing Office, 1986), p. 90.

nomic, and scientific gain. Tonight, I am directing NASA to develop a permanently manned space station and to do it within a decade.

A space station will permit quantum leaps in our research in science, communications, in metals, and in lifesaving medicines which could be manufactured only in space. We want our friends to help us meet these challenges and share in their benefits. NASA will invite other countries to participate so we can strengthen peace, build prosperity, and expand freedom for all who share our goals.

Just as the oceans opened up a new world for clipper ships and Yankee traders, space holds enormous potential for commerce today. The market for space transportation could surpass our capacity to develop it. Companies interested in putting payloads into space must have ready access to private sector launch services. The Department of Transportation will help an expendable launch services industry get off the ground. We'll soon implement a number of executive initiatives, develop proposals for ease regulatory constraints, and, with NASA's help, promote private sector investment in space. . . .

READING NO. 24

THE PRESIDENTIAL REPORT ON
THE *CHALLENGER* ACCIDENT[1]

The 28 January 1986 explosion of the Space Shuttle Challenger *and the ensuing investigation invited comparison with the
events that followed the launch-pad fire of the Apollo 204
spacecraft almost 19 years before that resulted in the deaths of
three astronauts. During the earlier accident a politically strong
administrator was at the helm of NASA; James E. Webb persuaded the White House to allow NASA to take the lead in the
accident investigation. That investigation was largely technical,
and it was sufficiently rigorous and critical to be seen as
credible. It resulted primarily in engineering changes; what
managerial changes Webb made as a result were surgical in
nature, lest the agency's entire management corps be cast into
confusion. In contrast, after the* Challenger *accident NASA's
internal investigation took a back seat to the work of a White
House-appointed commission, chaired by former Secretary of
State William P. Rogers. NASA was unable to seize the initiative
because, among other factors, its own top management was in
disarray. The report of the "Rogers Commission" was deliberate
and thorough, and, as this excerpt suggests, gave as much
emphasis to the accident's managerial as to its technical origins.*

γ   γ   γ

Pressures on the System

With the 1982 completion of the orbital flight test series,
NASA began a planned acceleration of the Space Shuttle launch
schedule. One early plan contemplated an eventual rate of a
mission a week, but realism forced several downward revisions.

[1]*Report of the Presidential Commission on the Space Shuttle Challenger Accident, Vol. I* (Washington, DC: Government Printing Office, 1986), pp. 164–177.

In 1985, NASA published a projection calling for an annual rate of 24 flights by 1990. Long before the Challenger accident, however, it was becoming obvious that even the modified goal of two flights a month was overambitious.

In establishing the schedule, NASA had not provided adequate resources for its attainment. As a result, the capabilities of the system were strained by the modest nine-mission rate of 1985, and the evidence suggests that NASA would not have been able to accomplish the 15 flights scheduled for 1986. These are the major conclusions of a Commission examination of the pressures and problems attendant upon the accelerated launch schedule.

On the same day that the initial orbital tests concluded—July 4, 1982—President Reagan announced a national policy to set the direction of the U.S. space program during the following decade. . . . From the inception of the Shuttle, NASA had been advertising a vehicle that would make space operations "routine and economical." The greater the annual number of flights, the greater the degree of routinization and economy, so heavy emphasis was placed on the schedule. However, the attempt to build up to 24 missions a year brought a number of difficulties, among them the compression of training schedules, the lack of spare parts, and the focusing of resources on near-term problems.

One effect of NASA's accelerated flight rate and the agency's determination to meet it was the dilution of the human and material resources that could be applied to any particular flight.

The part of the system responsible for turning the mission requirements and objectives into flight software, flight trajectory information and crew training materials was struggling to keep up with the flight rate in late 1985, and forecasts showed it would be unable to meet its milestones for 1986. It was falling behind because its resources were strained to the limit, strained by the flight rate itself and by the constant changes it was forced to respond to within that accelerating schedule. Compounding the problem was the fact that NASA had difficulty evolving from its single-flight focus to a system that could efficiently support the projected flight rate. It was slow in developing a hardware maintenance plan for its reusable fleet and slow in developing the capabilities that would allow it to handle the higher volume of work and training associated with the increased flight frequency.

Pressures developed because of the need to meet customer commitments, which translated into a requirement to launch a certain number of flights per year and to launch them on time. Such considerations may occasionally have obscured engineering concerns. Managers may have forgotten—partly because of past success, partly because of their own well-nurtured image of the program—that the Shuttle was still in a research and development phase. . . .

The sections that follow will discuss the pressures exerted on the system by the flight rate, the reluctance to relax the optimistic schedule, and the attempt to assume an operational status.

Planning of a Mission

The planning and preparation for a Space Shuttle flight require close coordination among those making the flight manifest, those designing the flight and the customers contracting NASA's services. The goals are to establish the manifest; define the objectives, constraints and capabilities of the mission; and translate those into hardware, software and flight procedures.

There are major program decision points in the development of every Shuttle flight. At each of these points, sometimes called freeze points, decisions are made that form the basis for further engineering and product development. The disciplines affected by these freeze points include integration hardware, engineering, crew timeline, flight design and crew training.

The first major freeze point is at launch minus 15 months. At that time the flight is officially defined: the launch date, Orbiter and major payloads are all specified, and initial design and engineering are begun based on this information.

The second major freeze point is at launch minus 7.7 months, the cargo integration hardware design. Orbiter vehicle configuration, flight design and software requirements are agreed to and specified. Further design and engineering can then proceed.

Another major freeze point is the flight planning and stowage review at launch minus five months. At that time, the crew activity timeline and the crew compartment configuration, which includes middeck payloads and payload specialist assignments, are established. Final design, engineering and training are based on these products.

Development of Flight Products

The "production process" begins by collecting all mission objectives, requirements and constraints specified by the payload and Space Shuttle communities at the milestones described above. That information is interpreted and assimilated as various groups generate products required for a Space Shuttle flight: trajectory data, consumables requirements, Orbiter flight software, Mission Control Center software and the crew activity plan, to name just a few.

Some of these activities can be done in parallel, but many are serial. Once a particular process has started, if a substantial change is made to the flight, not only does that process have to be started again, but the process that preceded it and supplied its date may also need to be repeated. If one group fails to meet its due date, the group that is next in the chain will start late. The delay then cascades through the system.

Were the elements of the system meeting their schedules? Although each group believed it had an adequate amount of time allotted to perform its function, the system as a whole was falling behind. An assessment of the system's overall performance is best made by studying the process at the end of the production chain: crew training. Analysis of training schedules for previous flights and projected training schedules for flights in the spring and summer of 1986 reveals a clear trend: less and less time was going to be available for crew members to accomplish their required training. . . .

The production system was disrupted by several factors including increased flight rate, lack of efficient production processing and manifest changes.

Changes in the Manifest

Each process in the production cycle is based on information agreed upon at one of the freeze points. If that information is later changed, the process may have to be repeated. The change could be a change in manifest or a change to the Orbiter hardware or software. The hardware and software changes in 1985 usually were mandatory changes; perhaps some of the manifest changes were not.

The changes in the manifest were caused by factors that fall into four general categories: hardware problems, customer requests, operational constraints and external factors. The significant changes made in 1985 are shown in the accompanying table. The following examples illustrate that a single proposed change can have extensive impact, not because the change itself is particularly difficult to accommodate (though it may be), but because each change necessitates four or five other changes. The cumulative effect can be substantial. . . .

When a change occurs, the program must choose a response and accept the consequences of that response. The options are usually either to maximize the benefit to the customer or to minimize the adverse impact on Space Shuttle operations. If the first option is selected, the consequences will include short-term and/or long-term effects. . . .

The effects of such changes in terms of budget, cost and manpower can be significant. In some cases, the allocation of additional resources allows the change to be accommodated with little or no impact to the overall schedule. In those cases, steps that need to be re-done can still be accomplished before their deadlines. The amount of additional resources required depends, of course, on the magnitude of the change and when the change occurs: early changes, those before the cargo integration review, have only a minimal impact; changes at launch minus five months (two months after the cargo integration review) can carry a major impact, increasing the required resources by approximately 30 percent. In the missions from 41-C to 51-L, only 60 percent of the major changes occurred before the cargo integration review. More than 20 percent occurred after launch minus five months and caused disruptive budget and manpower impacts.

Engineering flight products are generated under a contract that allows for increased expenditures to meet occasional high workloads. Even with this built-in flexibility, however, the requested changes occasionally saturate facilities and personnel capabilities. The strain on resources can be tremendous. For short periods of two to three months in mid-1985 and early 1986, facilities and personnel were being required to perform at roughly twice the budgeted flight rate.

If a change occurs late enough, it will have an impact on the serial processes. In these cases, additional resources will not alleviate the problem, and the effect of the change is absorbed by all downstream processes, and ultimately by the last element in the chain. In the case of the flight design and software reconfiguration process, that last element is crew training. In January, 1986, the forecasts indicated that crews on flights after 51-L would have significantly less time than desired to train for their flights. . . .

"Operational" Capabilities

For a long time during Shuttle development, the program focused on a single flight, the first Space Shuttle mission. When the program became "operational," flights came more frequently, and the same resources that had been applied to one flight had to be applied to several flights concurrently. Accomplishing the more pressing immediate requirements diverted attention from what was happening to the system as a whole. That appears to be one of the many telling differences between a "research and development" program and an "operational program." Some of the differences are philosophical, some are attitudinal and some are practical.

Elements within the Shuttle program tried to adapt their philosophy, their attitude and their requirements to the "operational era." But that era came suddenly, and in some cases, there had not been enough preparation for what "operational" might entail. For example, routine and regular post-flight maintenance and inspections are critical in an operational program; spare parts are critical to flight readiness in an operational fleet; and the software tools and training facilities developed during a test program may not be suitable for the high volume of work required in an operational environment. In many respects, the system was not prepared to meet an "operational" schedule.

As the Space Shuttle system matured, with numerous changes and compromises, a comprehensive set of requirements was developed to ensure the success of a mission. What evolved was a system in which the preflight processing, flight planning, flight control and flight training were accomplished with extreme care applied to every detail. This process checked and rechecked everything, and though it was both labor- and time-intensive, it was appropriate and necessary for a system still in the developmental phase. This process, however, was not capable of meeting the flight rate goals.

After the first series of flights, the system developed plans to accomplish what was required to support the fight rate. The challenge was to streamline the processes through automation, standardization, and centralized management, and to convert from the developmental phase to the mature system without a compromise in quality. It required that experts carefully analyze their areas to determine what could be standardized and automated, then take the time to do it.

But the increasing flight rate had priority—quality products had to be ready on time. Further, schedules and budgets for developing the needed facility improvements were not adequate. Only the time and resources left after supporting the flight schedule could be directed toward efforts to streamline and standardize. In 1985, NASA was attempting to develop the capabilities of a production system. But it was forced to do that while responding—with the same personnel—to a higher flight rate.

At the same time the flight rate was increasing, a variety of factors reduced the number of skilled personnel available to deal with it. These included retirements, hiring freezes, transfers to other programs like the Space Station and transitioning to a single contractor for operations support.

The flight rate did not appear to be based on assessment of available resources and capabilities and was not reduced to accommodate the capacity of the work force. For example, on January 1, 1986, a new contract took effect at Johnson that consolidated the entire contractor work force under a single company. This transition was another disturbance at a time when the work force needed to be performing at full capacity to meet the 1986 flight rate. In some important areas, a significant fraction of workers elected not to change contractors. This reduced the work force and its capabilities, and necessitated intensive training programs to qualify the new personnel. According to projections, the work force would not have been back to full capacity until the summer of 1986. This drain on a critical part of the system came just as NASA was beginning the most challenging phase of its flight schedule.

Similarly, at Kennedy the capabilities of the Shuttle processing and facilities support work force became increasingly strained as the Orbiter turnaround time decreased to accommodate the accelerated launch schedule. This factor has resulted in overtime percentages of almost 28 percent in some directorates. Numerous contract employees have worked 72 hours per week or longer and frequent 12-hours shifts. The potential implications of such overtime for safety were made apparent during the attempted launch of mission 61-C on January 6, 1986, when fatigue and shiftwork were cited as major contributing factors to a serious incident involving a liquid oxygen depletion that occurred less than five minutes before scheduled lift off. . . .

Responding to Challenges and Changes

Another obstacle in the path toward accommodation of a higher flight rate is NASA's legendary "can-do" attitude. The attitude that enabled the agency to put men on the moon and to build the Space Shuttle will not allow it to pass up an exciting challenge—even though accepting the challenge may drain resources from he more mundane (but necessary) aspects of the program.

A recent example is NASA's decision to perform a spectacular retrieval of two communications satellites whose upper-stage motors had failed to raise them to the proper geosynchronous orbit. NASA itself then proposed to the insurance companies who owned the failed satellites that the agency design a mission to rendezvous with them in turn and that an astronaut in a jet backpack fly over to escort the satellites into the Shuttle's payload bay for a return to Earth.

The mission generated considerable excitement within NASA and required a substantial effort to develop the necessary techniques, hardware and procedures. The mission was conceived, created, designed and accomplished within 10 months. The result, mission 51-A (November, 1984), was a resounding success, as both failed satellites were successfully returned to

Earth. The retrieval mission vividly demonstrated the service that astronauts and the Space Shuttle can perform.

Ten months after the first retrieval mission, NASA launched a mission to repair another communications satellite that had failed in low-Earth orbit. Again, the mission was developed and executed on relatively short notice and was resoundingly successful for both NASA and the satellite insurance industry.

The satellite retrieval missions were not isolated occurrences. Extraordinary efforts on NASA's part in developing and accomplishing missions will, and should, continue, but such efforts will be a substantial additional drain on resources. NASA cannot both accept the relatively spur-of-the-moment missions that its "can-do" attitude tends to generate and also maintain the planning and scheduling discipline required to operate as a "space truck" on a routine and cost-effective basis. As the flight rate increases, the cost in resources and the accompanying impact on future operations must be considered when infrequent but extraordinary efforts are undertaken. The system is still not sufficiently developed as a "production line" process in terms of planning or implementation procedures. It cannot routinely or even periodically accept major disruptions without considerable cost. NASA's attitude historically has reflected the position that "We can do anything," and while that may essentially be true, NASA's optimism must be tempered by the realization that it cannot do everything.

NASA has always taken a positive approach to problem solving and has not evolved to the point where its officials are willing to say they no longer have the resources to respond to proposed changes. . . .

It is important to determine how many flights can be accommodated, and accommodated safely. NASA must establish a realistic level of expectation, then approach it carefully. Mission schedules should be based on a realistic assessment of what NASA can do safely and well, not on what is possible with maximum effort. The ground rules must be established firmly, and then enforced.

The attitude is important, and the word operational can mislead. "Operational" should not imply any less commitment to quality or safety, nor a dilution of resources. The attitude should be, "We are going to fly high risk flights this year; every one is going to be a challenge, and every one is going to involve some risk, so we had better be careful in our approach to each." . . .

Findings

1. The capabilities of the system were stretched to the limit to support the flight rate in winter 1985/1986. Projections into the spring and summer of 1986 showed a clear trend; the system, as it existed, would have been unable to deliver crew training software for scheduled flights by the designated dates. The result would have been an unacceptable compression of the time available for the crews to accomplish their required training.

2. Spare parts are in critically short supply. The Shuttle program made a conscious decision to postpone spare parts procurements in favor of budget items of perceived higher priority. Lack of spare parts would likely have limited flight operations in 1986.

3. Stated manifesting policies are not enforced. Numerous late manifest changes (after the cargo integration review) have been made to both major payloads and minor payloads throughout the Shuttle program.

> Late changes to major payloads or program requirements can require extensive resources (money, manpower, facilities) to implement.

> If many late changes to "minor" payloads occur, resources are quickly absorbed.

> Payload specialists frequently were added to a flight well after announced deadlines.

> Late changes to a mission adversely affect the training and development of procedures for subsequent missions.

4. The scheduled flight rate did not accurately reflect the capabilities and resources.

> The flight rate was not reduced to accommodate periods of adjustment in the capacity of the work force. There was no margin in the system to accommodate unforeseen hardware problems.

> Resources were primarily directed toward supporting the flights and thus not enough were available to improve and expand facilities needed to support a higher flight rate.

5. Training simulators may be the limiting factor on the flight rate: the two current simulators cannot train crews for more than 12-15 flights per year.

6. When flights come in rapid succession, current requirements do not ensure that critical anomalies occurring during one flight are identified and addressed appropriately before the next flight.

READING NO. 25

PRINCIPAL RECOMMENDATIONS OF THE ADVISORY COMMISSION ON THE FUTURE OF THE U.S. SPACE PROGRAM[1]

Because of the difficulties NASA encountered in its major programs at the end of the 1980s, as well as the need periodically to review status and chart the course for the future, in 1990 President George Bush chartered an Advisory Committee on the Future of the U.S. Space Program under the leadership of Norman Augustine, chief executive officer of Martin Marietta. In the fall of the year Augustine submitted his commission's report, delineating the chief objectives of the agency and recommending several key actions. All of these related to the need to create a balanced space program—one that included human spaceflight, robotic probes, space science, applications, and exploration—within a tightly constrained budget.

γ   γ   γ

Principal Recommendations

This report offers specific recommendations pertaining to civil space goals and program content as well as suggestions relating to internal NASA management. These are summarized below in four primary groupings. In order to implement fully these recommendations and suggestions, the support of both the Executive Branch and Legislative Branch will be needed, and of NASA itself.

[1]*Report of the Advisory Committee on the Future of the U.S. Space Program* (Washington, DC: Government Printing Office, 1990), pp. 47–48.

Principal Recommendations Concerning Space Goals

It is recommended that the United States' future civil space program consist of a balanced set of five principal elements:

—a science program, which enjoys highest priority within the civil space program, and is maintained at or above the current fraction of the NASA budget (Recommendations 1 and 2);

—a Mission to Planet Earth (MTPE) focusing on environmental measurements (Recommendation 3);

—a Mission from Planet Earth (MFPE), with the long-term goal of human exploration of Mars, preceded by a modified Space Station which emphasizes life-sciences, an exploration base on the moon, and robotic precursors to Mars (Recommendations 4, 5, 6, and 7);

—a significantly expanded technology development activity, closely coupled to space mission objectives, with particular attention devoted to engines + a robust space transportation system (Recommendation 9).

Principal Recommendations Concerning Programs

With regard to program content, it is recommended that:

—the strategic plan for science currently under consideration be implemented (Recommendation 2);

—a revitalized technology plan be prepared with strong input from the mission offices, and that it be funded (Recommendation 8);

—Space Shuttle missions be phased over to a new unmanned (heavy-lift) launch vehicle except for mission where human involvement is essential or other critical national needs dictate (Recommendation 9);

—Space Station Freedom be revamped to emphasize life-sciences and human space operations, and include microgravity research as appropriate. It should be reconfigured to reduce cost and complexity; and the current 90-day time limit on redesign should be extended if a thorough reassessment is not possible in that period (Recommendation 6);

—a personnel module be provided, as planned, for emergency return from Space Station Freedom, and that initial provisions be made for two-way missions in the event of unavailability of the Space Shuttle (Recommendation 11).

Principal Recommendations Concerning Affordability

It is recommended that the NASA program be structured in scope so as not to exceed a funding profile containing approximately 10 percent real growth per year throughout the remainder of the decade and then remaining at that level, including but not limited to the following actions:

—redesign and reschedule the Space Station Freedom to reduce cost and complexity (Recommendation 6);

—defer or eliminate the planned purchase of another orbiter (Recommendation 10);

—place the Mission from Planet Earth on a "go-as-you-pay" basis, i.e., tailoring the schedule to match the availability of funds (Recommendation 5).

Principal Recommendations Concerning Management

With regard to management of the civil space program, it is recommended that:

—an Executive Committee of the Space Council be established which includes the Administrator of NASA (Recommendation 12);

—major reforms be made in the civil service regulations as they apply to specialty skills; or, if that is not possible, exemptions be granted to NASA for at least 10 percent of its employees to operate under a tailored personnel system; or, as a final alternative, that NASA begin selectively converting at least some of its centers into university-affiliated Federally Funded Research and Development Centers (Recommendations 14 and 15);

—NASA management review the mission of each center to consolidate and refocus centers of excellence in currently relevant fields with minimum overlap among centers (Recommendation 13).

It is considered by the Committee that the *internal* organization of any institution should be the province of, and at the discretion of, those bearing ultimate responsibility for the performance of that institution. . . .

—That the current headquarters structure be revamped, dis-establishing the positions of certain existing Associate Administrators . . .

—an exceptionally well-qualified independent cost analysis group be attached to headquarters with ultimate responsibility for all top-level cost estimating including cost estimates provided outside of NASA;

—a systems concept and analysis group reporting to the Administrator of NASA be established as a Federally Funded Research and Development Center;

—multi-center projects be avoided wherever possible, but when this is not practical, a strong and independent project office reporting to headquarters be established near the center having the principle share of the work for that project; and that this project office have a systems engineering staff and full budget authority (ideally industrial funding—i.e., funding allocations related specifically to end-goals).

In summary, we recommend:

1) Establishing the science program as the highest priority element of the civil space program, to be maintained at or above the current fraction of the budget.

2) Obtaining exclusions for a portion of NASA's employees from existing civil service rules or, failing that, beginning a gradual conversion of selected centers to Federally Funded Research and Development Centers affiliated with universities, using as a model the Jet Propulsion Laboratory.

3) Redesigning the Space Station Freedom to lessen complexity and reduce cost, taking whatever time may be required to do this thoroughly and innovatively.

4) Pursuing a Mission from Planet Earth as a complement to the Mission to Planet Earth, with the former having Mars as its very long-term goal—but relieved of schedule pressures and progressing according to the availability of funding.

5) Reducing our dependence on the Space Shuttle by phasing over to a new unmanned heavy lift launch vehicle for all but missions requiring human presence.

The Committee would be pleased to meet again in perhaps six months should the NASA Administrator so desire, in order to assist on the implementation process. In the meantime, NASA may wish to seek the assistance of its regular outside advisory group, the NASA Advisory Council, to provide independent and ongoing advice for implementing these findings.

Each of the recommendations herein is supported unanimously by the members of the Advisory Committee on the Future of the U.S. Space Program.

SELECTED BIBLIOGRAPHY AND SUGGESTIONS FOR FURTHER READING

The literature on the origins and development of the civil space program in the United States is uneven at best. The topic has captured some scholarly attention, but there is little that explores the broad social, cultural, economic, technological, and scientific influences of the space program. Existing investigation has largely been in three genres: (1) detailed historical scholarship on isolated aspects of the space program, most of which was underwritten by the National Aeronautics and Space Administration as official histories of specific programs, (2) journalistic works which are sometimes sensationalistic and always written for a popular audience without the scholar's attention to detail, and (3) policy studies prepared more to influence present decisions than to understand the space program's history. There is also a large number of works prepared for a popular audience which emphasize striking photographic images of space and contain all too often inane text.

There are several general overviews of the history of rocketry and space travel. Willy Ley, *Rockets, Missiles, and Space Travel* (New York: The Viking Press, 1961 ed.) and Eugene M. Emme, *A History of Space Flight* (New York: Holt, Rinehart and Winston, 1965), are outdated. Wernher von Braun, Frederick I. Ordway III, and Dave Dooling, *History of Rocketry and Space Travel* (New York: Thomas Y. Crowell Co., 1986 ed.), and Frederick I. Ordway III and Randy Lieberman, eds., *Blueprint for Space: Science Fiction to Science Fact* (Washington, DC: Smithsonian Institution Press, 1992), both provide well-illustrated introductions. Perhaps the most useful short history is Roger E. Bilstein, *Orders of Magnitude: A History of the NACA and NASA, 1915–1990* (NASA SP-4406, 1989), but it is not readily available except in government repositories.

The technology of rocketry has been explored in several uneven publications. Perhaps the best overall introduction to the subject in an international context can be found in Frank H. Winter, *Rockets into Space* (Cambridge, MA: Harvard University Press, 1990). Eugene M. Emme, ed. *The History of Rocket*

Technology (Detroit, MI: Wayne State University Press, 1964), and Stanley M. Ulanoff, *Illustrated Guide to U.S. Missiles and Rockets* (Garden City, NY: Doubleday and Co., 1962), are more detailed, but they are a generation old. Most of the U.S. civil space program's early launch vehicles came from military developments during and after World War II. Gregory P. Kennedy, *Vengeance Weapon 2: The V-2 Guided Missile* (Washington, DC: Smithsonian Institution Press, 1983) is a survey of the development and use of the first true ballistic missile. David H. DeVorkin, *Science with a Vengeance: How the Military Created the US Space Sciences After World War II* (New York: Springer-Verlag, 1992), is a lively account of the early post-war marriage of rocket technology and science to study the universe beyond Earth. Edmund Beard, *Developing the ICBM: A Study in Bureaucratic Politics* (New York: Columbia University Press, 1976), and Jacob Neufeld, *Ballistic Missiles in the United States Air Force, 1945–1960* (Washington, DC: Office of Air Force History, 1990), are political and institutional accounts of the development of these new weapons. John L. Sloop, *Liquid Hydrogen as a Propulsion Fuel, 1945–1959* (NASA SP-4404, 1978), and Stephen E. Doyle, ed., *History of Liquid Rocket Engine Development in the United States, 1955–1980* (San Diego, CA: Univelt, AAS History Series, 1992), are specialized studies of technological development. David Baker, *The Rocket: The History and Development of Rocket and Missile Technology* (New York: Crown Books, 1978), is a large-format, profusely illustrated overview.

Penetrating biographical studies of space program pioneers are in short supply. The three great rocketry pioneers of the early twentieth century are profiled with varying degrees of inadequacy in Arkady Kosmodemyansky, *Konstantin Tsiolkovsky* (Moscow, USSR: Nauka, 1985); Helen B. Walters, *Hermann Oberth: Father of Space Travel* (New York: Macmillan, 1962); and Milton Lehman, *This High Man* (New York: Farrar, Straus, 1963), but all three deserve modern scholarly biographies. Wernher von Braun has only been portrayed in Erik Bergaust, *Wernher von Braun* (Washington, DC: National Space Institute, 1976), and Ernst Stulinger and Frederick I. Ordway III, *Wernher von Braun: Crusader for Space*, 2 vols. (Melbourne, FL: Krieger Pub. Co., 1993), but there is room for serious historical investigation. Some biographical articles are contained in Frederick C. Durant III and George S. James, eds., *First Steps Toward Space: Proceedings of the First and Second History Symposia of the International Academy of Astronautics* (Washington, DC: Smithsonian Institution Press, 1974), and the several volumes of the *History of Rocketry and Astronautics*, sponsored by the International Academy of Astronautics and published by Univelt, Inc., in San Diego, CA.

In contrast, the history of space science has received considerable attention from historians. A fine short introduction to the history of astronomy is Anton Pannekoek, *A History of Astronomy* (New York: Dover, 1993 ed.), while a comprehensive work is Owen T Gingerich, gen. ed., *The Cambridge General History of Astronomy* (New York: Cambridge University Press, 1984–), 3 vols. to date. A good introduction to the physical science of this effort can be found in Lloyd Motz and Jefferson Hane Weaver, *The Story of Physics* (New York: Avon Books, 1989). The contributions of space science to the understanding of the universe have been discussed in Bevan M. French and Stephen P. Maran, eds., *A Meeting with the Universe: Science Discoveries from the Space Program* (NASA EP-177, 1981); and Paul A. Hanle and Von Del Chamberlain, eds., *Space Science Comes of Age: Perspectives in the History of the Space Sciences* (Washington, DC: Smith Institution Press, 1981). Joseph N. Tatarewicz, *Space Technology and Planetary Astronomy* (Indiana University Press, 1990); Robert W. Smith, *The Space Telescope: A Study of NASA, Science, Technology, and Politics* (New York: Cambridge University Press, 1989); and Karl Hufbauer, *Exploring the Sun: Solar Science Since Galileo* (Johns Hopkins University Press, 1991), provide considerable information on the development of scientific studies using space technology in the twentieth century.

One of the most important events relating to the development of the U.S. space program was the Sputnik crisis, inaugurated by the Soviet launch of a scientific satellite in 1957. A standard history of this event and the early years of NASA, emphasizing international rivalries between the United States and the Soviet Union, is Walter A. McDougall, . . . *The Heavens and the Earth: A Political History of the Space Age* (New York: Basic Books, 1985). Also on this subject see Robert A. Divine, *The*

Sputnik Crisis (New York: Oxford University Press, 1993); Rip Bulkeley, *The Sputniks Crisis and Early United States Space Policy: A Critique of the Historiography of Space* (Bloomington: Indiana University Press, 1991); and Robert L. Rosholt, *An Administrative History of NASA, 1958–1963* (NASA SP-4101, 1966). Two valuable memoirs of this era are James R. Killian, Jr., *Sputnik, Scientists, and Eisenhower: A Memoir of the First Special Assistant to the President for Science and Technology* (Cambridge: MIT Press, 1977); and T. Keith Glennan, *The Birth of NASA: The Diary of T. Keith Glennan* (NASA SP-4105, 1993). Constance McLaughlin Green and Milton Lomask, *Vanguard: A History* (Washington, DC: Smithsonian Institution Press, 1971), deals with the U.S.'s pioneering space flight program in response to the Soviet Union.

Project Apollo changed the face of the U.S. space program, and it has received attention from several historians. John M. Logsdon, *The Decision to Go to the Moon: Project Apollo and the National Interest* (Cambridge: MIT Press, 1970), is still the standard history of the political decision. Charles Murray and Catherine Bly Cox, *Apollo: The Race to the Moon* (New York: Simon and Schuster, 1989), provide an excellent overview of a complex subject. A well-written account of the efforts to build the Saturn launch vehicle by von Braun and his colleagues can be found in Frederick I. Ordway III and Mitchell R. Sharpe, *The Rocket Team* (New York: Crowell, 1979). NASA official histories of the subject include: Charles D. Benson and William Barnaby Faherty, *Moonport: A History of Apollo Launch Facilities and Operations* (NASA SP-4204, 1978); Courtney G. Brooks, James M. Grimwood, and Loyd S. Swenson, Jr., *Chariots for Apollo: A History of Manned Lunar Spacecraft* (NASA SP-4205, 1979); Roger E. Bilstein, *Stages to Saturn: A Technological History of the Apollo/Saturn Launch Vehicles* (NASA SP-4206, 1980); Arnold S. Levine, *Managing NASA in the Apollo Era* (NASA SP-4102, 1982); W. David Compton, *Where No Man Has Gone Before: A History of Apollo Lunar Exploration Missions* (NASA SP-4214, 1989); and Sylvia D. Fries, *NASA Engineers and the Age of Apollo* (NASA SP-4104, 1992). The two projects that led up to Apollo are ably documented in Loyd S. Swenson, Jr., James M. Grimwood, and Charles C. Alexander, *This New Ocean: A History of Project Mercury* (Washington, DC: NASA SP-4201, 1966); and Barton C. Hacker and James M. Grimwood, *On Shoulders of Titans: A History of Project Gemini* (NASA SP-4203, 1977). A journalistic account of what Apollo meant can be found in *Of a Fire on the Moon* (Boston: Little, Brown, 1969), by Norman Mailer.

Some of the space projects of the 1970s have been profiled in individual books, however, most of these have appeared as a result of the efforts of NASA to document its history. W. David Compton and Charles D. Benson, *Living and Working in Space: A History of Skylab* (NASA SP-4208, 1983), deals with the orbital laboratory; while Edward Clinton Ezell and Linda Neuman Ezell, *The Partnership: A History of the Apollo-Soyuz Test Project* (NASA SP-4209, 1978), is an in-depth study of this "diplomatic mission." On Project Viking's survey of Mars see, Edward Clinton Ezell and Linda Neuman Ezell, *On Mars: Exploration of the Red Planet, 1958–1978* (NASA SP-4212, 1984). Applications for space technology has received considerable attention of late. The history of Landsat is ably documented in Pamela E. Mack, *Viewing the Earth: The Social Construction of the Landsat Satellite System* (Cambridge: MIT Press, 1990). Frederik Nebeker, *Calculating the Weather: Meteorology in the 20th Century* (Bloomington: Indiana University Press, 1993), includes information on satellite activities.

The largest NASA projects, the Space Shuttle and the Space Station *Freedom*, have not received much historical attention. There is no satisfactory one-volume history of the shuttle, despite the project's longevity. The best way to come to grips with its origin and development is to review the historiographical literature contained in Roger D. Launius and Aaron K. Gillette, comps., *The Space Shuttle: An Annotated Bibliography* (NASA, Studies in Aerospace History, No. 1, 1992). The only substantive work on the Space Station is Howard E. McCurdy, *The Space Station Decision: Incremental Politics and Technological Choice* (Johns Hopkins University Press, 1990), although a full-fledged project history is underway by NASA.

Finally, an important institutional culture study of NASA's evolution from origins to the 1990s, in the process commenting on many of the issues affecting the agency in the last decade of the twentieth century, is Howard E. McCurdy, *Inside NASA: High Technology and Organizational Change in the U.S. Space Program* (Johns Hopkins University Press, 1993).

INDEX

271